Prehospital Transport and Whole-Body Vibration

Prehospital Transport and Whole-Body Vibration

Salam Rahmatalla
Professor, Civil and Environmental Engineering,
Professor, Biomedical Engineering,
The University of Iowa, Iowa City, IA, United States

ACADEMIC PRESS

An imprint of Elsevier

ELSEVIER

Academic Press is an imprint of Elsevier
125 London Wall, London EC2Y 5AS, United Kingdom
525 B Street, Suite 1650, San Diego, CA 92101, United States
50 Hampshire Street, 5th Floor, Cambridge, MA 02139, United States
The Boulevard, Langford Lane, Kidlington, Oxford OX5 1GB, United Kingdom

Notices
Knowledge and best practice in this field are constantly changing. As new research and experience broaden our understanding, changes in research methods, professional practices, or medical treatment may become necessary.

Practitioners and researchers must always rely on their own experience and knowledge in evaluating and using any information, methods, compounds, or experiments described herein. In using such information or methods they should be mindful of their own safety and the safety of others, including parties for whom they have a professional responsibility.

To the fullest extent of the law, neither the Publisher nor the authors, contributors, or editors, assume any liability for any injury and/or damage to persons or property as a matter of products liability, negligence or otherwise, or from any use or operation of any methods, products, instructions, or ideas contained in the material herein.

British Library Cataloguing-in-Publication Data
A catalogue record for this book is available from the British Library

Library of Congress Cataloging-in-Publication Data
A catalog record for this book is available from the Library of Congress

ISBN: 978-0-323-90103-1

For Information on all Academic Press publications
visit our website at https://www.elsevier.com/books-and-journals

Publisher: Stacy Masucci
Acquisitions Editor: Elizabeth Brown
Editorial Project Manager: Pat Gonzalez
Production Project Manager: Punithavathy Govindaradjane
Cover Designer: Matthew Limbert

Typeset by MPS Limited, Chennai, India

Working together
to grow libraries in
developing countries

www.elsevier.com • www.bookaid.org

Contents

Preface

There are many situations in which humans are transported in supine positions and exposed to whole-body vibration (WBV). Many of these occur in air, sea, and ground vehicles en route to a hospital. The dynamics coupling between the human body and the transport vehicle can generate a complex environment. On one hand, the vehicle and transport system comprises many mechanical components with different material properties that can move unpredictably when exposed to random vibration. On the other hand, the human body is a very complicated, active biomechanical system that can generate unpredictable voluntary and involuntary movements during vibration. Combining and investigating these two systems during transport can be very challenging and requires multidisciplinary effort from people in the fields of engineering, emergency medicine, ergonomics, and human factors. Therefore this book is designed to provide to readers from a variety of disciplines background information, basic concepts, and real-life examples of WBV during transportation. It introduces lab and field case studies and discusses the different types of measurements and analyses being used in the WBV literature.

The transport of supine humans under random vibration can lead to severe involuntary motions of body segments, which can generate discomfort, pain, and secondary injuries, especially when the transported human is a patient who has trauma or suspected spinal cord injury. In light of this, immobilization systems are traditionally used during medical transport to secure patients and limit their movements to reduce unwanted motion. While immobilization systems have been used for a long time, there is an ongoing debate about the advantages and disadvantages of using them in the theater. People on one side say that immobilization systems play an important role and encourage their use, while those on the other side of the spectrum say the systems take critical prehospital time to implement, cause unnecessary discomfort to patients, and could potentially harm the patient. Ongoing research in this area is illuminating the pros and cons of different transport systems and will eventually provide enough evidence for emergency responders to determine best practices for patient transport. Research will also be required to develop transport systems that are more efficient and provide better support for patients. In order to achieve best practice, medical and technical resources should be combined and collaborative works should be conducted to truly understand the statistical and medical significance of using immobilization systems.

While there are many transport systems in use, their effectiveness, advantages, and disadvantages are still under debate. This book provides a background on how vibration affects human response and will help technical, medical, and ergonomics professionals evaluate research and current practice. It is hoped that this information will be used to generate safe transport environments for injured patients. The background information outlined in this book includes the introduction of different measurement systems that can be used in the area of WBV; the explanation of different mathematical and statistical terms and metrics that are used to quantify human response to vibration; the effect of human anthropometry, vibration magnitude, and immobilization systems on the involuntary motions that can occur to humans during transport; and the perception and quantification of the discomfort that can result during transport. The book also sheds light on the ongoing debate about the necessity of using immobilization systems for people with spinal injuries. It presents a summary of the research that has been done in the area of supine human response to WBV with a specific emphasis on prehospital transport. The majority of the chapters include case studies where concepts are presented and tested in more detail.

Chapter 1, Fundamentals of Motion and Biomechanics, presents basic concepts in motion kinematics and kinetics, different types of vibrations, indices used for the characterization of random vibrations, basic statistics and data analysis, and an introduction to human modeling in a WBV environment.

Chapter 2, Measurement of Human Response to Vibration, presents measurement techniques for capturing human responses. It covers sensors including marker-based passive systems, accelerometers, and inertial systems. Techniques on how to attach these systems to the human body, how to collect vibration data, and how to analyze the data are introduced.

Chapter 3, Biodynamics of Supine Humans Subjected to Vibration and Shocks, introduces the biodynamics, or response, of a supine human under WBV. It details how the human body responds and moves when the vibration intensity increases or when the frequency, or number of cycles per second, changes. The chapter also includes the effect of human gender, body mass, and anthropometry on human response during transport. The characteristics and roles of the different components of immobilization systems, including spine boards, neck collars, and straps, on the motion of the different segments of the human body, are also introduced.

Chapter 4, Discomfort in Whole-Body Vibration, presents research on discomfort during transport and its relationship with the severity of WBV. This includes materials related to supine positions, representing the patients, and seated positions, representing the medics or emergency responders.

Chapter 5, Justification and Efficacy of Prehospital Immobilization Systems, introduces human response to vibration during prehospital transport and illustrates how different immobilization systems restrict and reduce body movement during transport; differences in how the body is supported and immobilized can produce different effects. This chapter presents a summary of the work that has been presented in the literature, including the pros and cons of using immobilization systems for patients with potential spinal cord injury.

To ensure that this book is accessible and easy to read for readers from a variety of backgrounds, including engineers, technicians, medics, emergency medicine doctors, and ergonomists, it provides basic concepts in motion, vibration, and human response to WBV. More consideration will be given to physical meanings and applications and less to mathematical formulations; however, basic mathematical formulations are provided. References will direct interested readers to original sources for mathematical background and details. Many publications, including a handful of books, have been written about human response to vibration. This book will focus on supine-human response to vibration because of its relevance to prehospital transport.

Acknowledgments

I would like to say that this book could not have been possible without the contributions of my graduate students (Jonathan DeShaw, John Meusch, Guandong Qiao, Yang Wang, Ye Liu, Rosiland Heckman, and Eric Frick) and my research coauthors who have joined me in this journey for the last 15 years.

I would like to thank Caterpillar Inc. and Battelle Memorial Institute for sponsoring some of the research presented in this book.

I am very grateful to Ms. Melanie Laverman for her editorial support in the book chapters.

I would like to thank my wife Deema for reading the chapters and providing her medical perspective.

My warmest thanks to my daughter Tia for her patience and support.

Chapter 1

Fundamentals of motion and biomechanics

1.1 Introduction

The human body is a complex system that comprises many segments, namely, the head, neck, chest, abdomen, pelvis, arms, and legs, connected by muscles, tendons, ligaments, soft tissues, and blood vessels, among other things. On the other hand, vibration or shaking is generally a random and unpredictable motion with characteristics that can generate different types of forces that, when combined with sudden impacts (shocks), can be severe. Therefore putting humans in a vibration environment and studying their response is considered a very challenging task. Due to the complexity of the subject matter, the goal here is to present basic concepts to readers from different backgrounds. Readers with engineering backgrounds may skip the majority of the materials presented in the chapter; however, the materials present a summary of basic concepts in motion, statistics, and vibration, and readers may find it a good opportunity to review or learn these concepts before reading the rest of the chapters.

This chapter introduces fundamental concepts in motion and vibration with an emphasis on human response to vibration. The goal of this chapter is to familiarize the reader with the concepts of motion and forces and their effects on different systems, including the human body. In order to reach the point where the reader can understand the effect of vibration on the human body, a step-by-step process is introduced in this chapter to help readers with different backgrounds understand these concepts. This chapter also introduces some fundamental metrics in statistics and vibration that will help the reader understand the materials in the following chapters.

The chapter starts with a description of the motion of one particle without considering the effect of its mass or the forces applied to it (this is called kinematics of motion). Although a particle represents an idealized form of objects because it does not have dimensions, Newton's second law, which is considered the fundamental law of motion, is based on the concept of particles. This chapter introduces the concept of a particle to describe motion in terms of position, displacement, velocity, and acceleration. The need for establishing a reference coordinate system as an essential part of defining the motion of any object is also explained. Because the motion of any object can have magnitude

Prehospital Transport and Whole-Body Vibration. DOI: https://doi.org/10.1016/B978-0-323-90103-1.00002-4

and direction, for example, a velocity of 30 km/h north, the concept of vectors is introduced. As motion variables can change with time, their magnitude and direction can also change, so the chapter introduces schemes for vector differentiation and integration. The magnitude and direction of forces that can result in different forms based on the coordinate system used are then introduced, and the concept of a particle is extended to an object, where dimensions become important. With objects, the motion becomes more complicated because objects can not only translate like particles, but they can also rotate. Newton's second law is then extended to deal with objects, and examples illustrate the implementation of the established formulas.

The chapter then introduces the concept of vibration as well as different types of vibration. These include simple forms of deterministic free, damped, and forced systems. As with particles and motions, the introduction of simple vibratory systems with a single mass and one degree of freedom should help readers understand what can happen to complicated systems like the human body. While random vibration is mostly the norm in real-life applications, including transport, different fundamental concepts and metrics are introduced and explained, including the definition of transfer functions that relate the output motion to the input motion of the system. Understanding these concepts will help readers quantify and assess vibrations in the environments where they work. More detailed models are presented toward the end of the chapter to illustrate the degree of complexity that can be involved in predicting the involuntary motion of humans inside a vibration environment.

1.2   Basic vector algebra

Scalars, such as mass, are quantities that can be described by a single number with standardized units such as kilograms (kg) in the SI system, pound mass (lbm) in the British system, and *slug* in the United States customary system. Vectors, on the other hand, are mathematical representations of any physical property that has a magnitude and a direction. Velocity and acceleration are vectors because they are described by their magnitudes and directions; for example, a wind with a velocity of 30 kilometers per hour (km/h) in the northeast direction will have a different effect than a 30 km/h wind in the northwest direction.

Vectors are normally presented with arrows. The head of the arrow points in the direction of the vector, and the length of the arrow reflects its length or magnitude. Fig. 1.1A shows a schematic representation of a vector. Normally, a vector is described within a coordinate system, and a Cartesian coordinate system is used in many applications. The Cartesian coordinate system has three perpendicular, or orthogonal, axes, representing the space surrounding us or the space that we live in. They are normally represented as x, y, and z using the right-hand rule description. In the right-hand rule (Fig. 1.1B), the thumb of the right-hand points to one axis, such as the z-axis, and the curled fingers of the right hand represent a rotation starting from the x-axis and pointing to the y-axis.

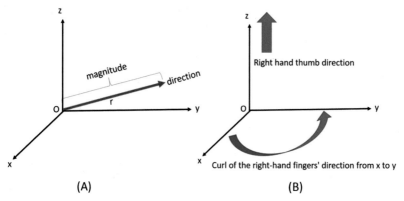

(A)   (B)

FIGURE 1.1   (A) Schematic representation of a vector showing its magnitude and direction;
(B) the right-hand rule between orthogonal axes.

Any vector in the Cartesian coordinate system can be represented by its
components along the axes (x, y, and z). The x, y, and z components are independent, meaning that analysis can be conducted in each axis direction independent of what is happening in other axes. The collection of these components at
any location will result in the vector. It should be noted that the vector magnitude will be the same regardless of the use of any coordinate system.

The vector **r** in Fig. 1.2 shows the position of a point P with coordinates
4, 2, and 6. In this representation, the vector has 4 units in the x-direction, 2
units in the y-direction, and 6 units in the z-direction.

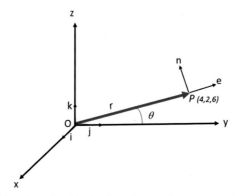

FIGURE 1.2   Position of a point P in the Cartesian coordinate system, showing a unit vector
(**e**) in the direction of the vector (**r**) and a unit vector perpendicular to it (**n**).

This vector can be represented using the following mathematical form:

$$\mathbf{r} = 4i + 2j + 6k$$

where i, j, and k are unit vectors. A unit vector is a vector that has a length
of one unit. The unit vectors i, j, and k are always pointing in the directions

of x, y, and z, respectively. Other types of unit vectors can be developed to point in any direction. Vectors can be represented in different forms depending on the coordinate systems they are used in. For example, in the polar (two-dimensional) or cylindrical (if the z-axis is included to become three-dimensional) coordinate system, any vector can be represented by its length and the angle it makes with the coordinate axes.

In the case shown in Fig. 1.2, point P can also be described by the coordinates r and θ, indicating the distance that P is from the center or origin of the coordinate system and the angle it is making with the x-axis. The vector $\mathbf{r}$ can be written in terms of its magnitude r and direction $\mathbf{e}$ as follows:

$$\mathbf{r} = r\mathbf{e} \tag{1.1}$$

where the magnitude or the length of the vector $\mathbf{r}$,

$$r = \sqrt{x^2 + y^2 + z^2} = \sqrt{4^2 + 2^2 + 6^2} = 7.483$$

and $\mathbf{e}$ is a unit vector pointing in the direction of $\mathbf{r}$ and can be calculated as follows:

$$\mathbf{e} = \frac{\text{The vector}}{\text{Magnitude of the vector}} = \frac{\mathbf{r}}{r} = \frac{4i + 2j + 6k}{7.483} = 0.535i + 0.267j + 0.802k$$

It can be seen that the magnitude or length of the unit vector $\mathbf{e}$ is

$$|\mathbf{e}| = e = \sqrt{0.545^2 + 0.267^2 + 0.802^2} = 1$$

Let us take another example and look at the meaning of the unit vector $\mathbf{e}$ in a plane (x, y).

Let us have

$$\mathbf{r} = r\mathbf{e} = 4i + 3j$$

where

$$r = \sqrt{4^2 + 3^2} = 5$$

and

$$\mathbf{e} = \frac{\mathbf{r}}{r} = \frac{4i + 3j}{5} = 0.8i + 0.6j$$

As shown in Fig. 1.3, the component of the unit vector $\mathbf{e}$ in the x-direction (0.8) represents the cosine of the angle it is making with the x-axis, that is,

$$\cos\theta = 0.8 \Rightarrow \theta = \cos^{-1}(0.8) = 36.86°,$$

and the component of the unit vector $\mathbf{e}$ in the y-direction (0.6) represents the cosine of the angle it is making with the y-axis, that is,

$$\cos\alpha = 0.6 \Rightarrow \alpha = \cos^{-1}(0.6) = 53.13°$$

So the components of the unit vector are the cosines of the angle it is making with the coordinate system.

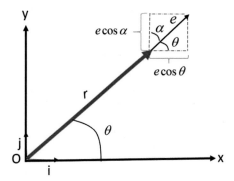

FIGURE 1.3 The component of the unit vector **e** in the *x*- and *y*-directions as a function of the cosine of the angles.

1.2.1 Vector addition and subtraction

Vectors like scalar quantities can be added and subtracted as shown in the following example. If

$$\mathbf{v}_1 = 4i + 2j + 6k$$

and

$$\mathbf{v}_2 = 2i + j + 2k$$

then

$$\mathbf{v}_1 + \mathbf{v}_2 = (4i + 2j + 6k) + (2i + j + 2k) = 6i + 3j + 8k$$

and

$$\mathbf{v}_1 - \mathbf{v}_2 = (4i + 2j + 6k) - (2i + j + 2k) = 2i + j + 4k$$

As shown in the example, the components in each similar direction (*i*, *j*, and *k*) are added together.

1.2.2 Vector multiplication

The multiplication of vectors can be done in different ways, including inner or dot and cross product. In the dot product of two vectors (Fig. 1.4), the result is a scalar. That is why the dot product is sometimes called a scalar product. The mathematical formula for the dot product is as follows:

$$\mathbf{v}_1 \cdot \mathbf{v}_2 = v_1 v_2 \cos\theta \tag{1.2}$$

where θ is the angle between the two vectors, and v_1 and v_2 are the magnitudes of the vectors.

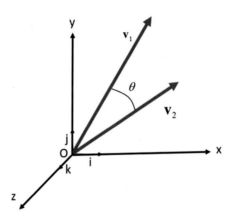

FIGURE 1.4   Vectors in the Cartesian system.

When the two vectors become perpendicular to each other, that is, when $\theta = 90°$, the dot product result will be zero because $\cos 90° = 0$. The dot product can also be done by multiplying each component of $\mathbf{v}_1$ by its corresponding component in each direction in $\mathbf{v}_2$ as follows.
Let

$$\mathbf{v}_1 = 4i + 2j + 6k$$

and

$$\mathbf{v}_2 = 2i + j + 2k$$

then

$$\mathbf{v}_1 \cdot \mathbf{v}_2 = (4i + 2j + 6k) \cdot (2i + j + 2k) = ((4)(2) + (2)(1) + (6)(2)) = 22$$

When the dot product is conducted on the vector itself, the result of this multiplication represents the norm or the length of the vector as follows:

$$\|\mathbf{v}_1\| = \sqrt{\mathbf{v}_1 . \mathbf{v}_1} = \sqrt{(4i + 2j + 6k) \cdot (4i + 2j + 6k)} = \sqrt{56} = 7.483 \qquad (1.3)$$

$$\|\mathbf{v}_2\| = \sqrt{\mathbf{v}_2 \cdot \mathbf{v}_2} = \sqrt{(2i + j + 2k) \cdot (2i + j + 2k)} = \sqrt{9} = 3$$

The dot product can also be used to determine the angle between two vectors $\mathbf{v}_1$ and $\mathbf{v}_2$ as follows:

$$\cos\theta = \frac{\mathbf{v}_1 \cdot \mathbf{v}_2}{|\mathbf{v}_1||\mathbf{v}_2|} = \frac{22}{7.483 \times 3} \to \theta = \cos^{-1}(0.98) = 11°.48 \qquad (1.4)$$

This relationship is sometimes used to check whether two vectors are perpendicular to each other; in this case, if the dot product between the two vectors is zero, that is, $\theta = 90°$, then the vectors are orthogonal, or perpendicular, to each other. One implication of this is the orthogonality of the unit vectors in the Cartesian coordinate system (Fig. 1.2), where $\mathbf{i} \cdot \mathbf{j} = \mathbf{j} \cdot \mathbf{i} = \mathbf{i} \cdot \mathbf{k} = \mathbf{j} \cdot \mathbf{k} = 0$.

1.2.3   Projection of a vector in a certain direction

The dot product can be used to project a vector in any desired direction. This can be done by conducting a dot product between the vector with a unit vector in the intended projected direction.

Example:

The vector projection ($\bar{\mathbf{v}}$) of $\mathbf{v} = 4i + 2j + 6k$ in the direction of another vector $\mathbf{a} = 2i + j + 2k$ can be done as follows.

1. Find   a   unit   vector   in   the   direction   of   the   vector
   $$\mathbf{a} \Rightarrow e = (\mathbf{a}/a) = \left(2i + j + 2k / \sqrt{2^2 + 1^2 + 2^2}\right) = (2i/3) + (j/3) + (2k/3).$$

2. Find the magnitude of the projection (d) of $\mathbf{v}$ on the vector $\mathbf{a}$ (Fig. 1.5).

$$d = \mathbf{v} \cdot \mathbf{e} = (4i + 2j + 6k).\left(\frac{2}{3}i + \frac{1}{3}j + \frac{2}{3}k\right) = (4)\left(\frac{2}{3}\right) + (2)\left(\frac{1}{3}\right) + (6)\left(\frac{2}{3}\right) = 7.333$$

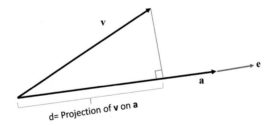

FIGURE 1.5   Projection of a vector $\mathbf{v}$ in the direction of another vector $\mathbf{a}$.

3. Multiply the magnitude of projection d in step 2 by $\mathbf{e}$ to create a vector $\bar{\mathbf{v}}$ in the direction of $\mathbf{a}$.

$$\bar{\mathbf{v}} = d\mathbf{e} = 7.333\left(\frac{2}{3}i + \frac{1}{3}j + \frac{2}{3}k\right) = 4.889i + 2.444j + 4.889k$$

1.2.4   Geometric representation of vectors

One interesting representation of a vector is the geometrical representation. A main advantage of this representation is the ability to represent a vector in the n-dimensional space, with the vector having many components in many directions, for applications that cannot be represented by the Cartesian space or coordinate system, which only has three independent axes (x, y, and z).

In the geometrical form, the vector can be written as $\mathbf{a} = [1\ 0\ 1]$ or $\mathbf{b} = [2\ 1\ 3\ 0\ \text{-}3\ 5, \ldots]$.

The vector $\mathbf{a}^T$ is the transpose of the vector $\mathbf{a}$. If $\mathbf{a}$ is a row vector, then $\mathbf{a}^T$ is a column vector and vice versa. The dot product of a vector by itself can be written as $\mathbf{a}^T \cdot \mathbf{a}$.

1.2.5 Cross product

While the dot product of two vectors results in a scalar or a number quantity, the cross product or multiplication between two vectors $\mathbf{a} = \begin{bmatrix} 1 & 0 & 1 \end{bmatrix}$ and $\mathbf{b} = \begin{bmatrix} 2 & 1 & 3 \end{bmatrix}$ gives a vector ($\mathbf{c}$) that is perpendicular to the plane containing the two vectors (Fig. 1.6). The cross product follows the right-hand rule, where the curled fingers of the right hand pass through the two vectors, and the thumb of the right hand points to the resulting vector of the cross product.

$$\mathbf{c} = \mathbf{a} \times \mathbf{b} = \begin{bmatrix} a_1 & a_2 & a_3 \end{bmatrix}\begin{bmatrix} b_1 & b_2 & b_3 \end{bmatrix} = \begin{bmatrix} a_2 b_3 - a_3 b_2 & a_3 b_1 - a_1 b_3 & a_1 b_2 - a_2 b_1 \end{bmatrix}$$
(1.5)

$$\mathbf{c} = \begin{bmatrix} c_1 & c_2 & c_3 \end{bmatrix} = \begin{bmatrix} 0 \times 3 - 1 \times 1 & 1 \times 2 - 1 \times 3 & 1 \times 1 - 0 \times 2 \end{bmatrix} = \begin{bmatrix} -1 & -1 & 1 \end{bmatrix}$$

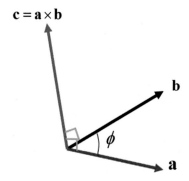

FIGURE 1.6 Cross product between two vectors.

The cross product has a length or a magnitude that can be calculated as follows:

$$c = ab\sin\phi \qquad (1.6)$$

where c is the magnitude of $\mathbf{c}$, a is the magnitude of $\mathbf{a}$, and b is the magnitude of $\mathbf{b}$, and ϕ is the angle between $\mathbf{a}$ and $\mathbf{b}$.

For $\phi = 30°$,

$$c = ab\sin\phi = \sqrt{1^2 + 0^2 + 1^2} \times \sqrt{2^2 + 1^2 + 3^2} \times \sin 30 = 2.645$$

If **a** and **b** have the same or opposite directions, their cross product is zero because

$$\sin\phi = \sin(0) = \sin(180) = 0.$$

The cross-product multiplication of the Cartesian unit vectors (Fig. 1.4) follows the right-hand rule, given as follows:

$$\mathbf{i} \times \mathbf{j} = \mathbf{k}, \mathbf{j} \times \mathbf{k} = \mathbf{i}, \mathbf{k} \times \mathbf{i} = \mathbf{j}, \mathbf{i} \times \mathbf{k} = -\mathbf{j}, \mathbf{k} \times \mathbf{j} = -\mathbf{i}, \text{ and } \mathbf{j} \times \mathbf{i} = -\mathbf{k}.$$

As can be seen, there is a minus sign if the cross product is conducted in the opposite order, that is,

$$\mathbf{i} \times \mathbf{j} = -(\mathbf{j} \times \mathbf{i}), \mathbf{j} \times \mathbf{k} = -(\mathbf{k} \times \mathbf{j}), \mathbf{k} \times \mathbf{i} = -(\mathbf{i} \times \mathbf{k})$$

1.2.6   Vector calculus

The derivative of a scalar quantity such as the mass represents the change in mass with respect to another parameter. For example, the derivative of the mass with respect to time is the rate of change of a mass with respect to time, or how much the mass is changed with time. It is normally represented as

$$\frac{dm}{dt} = \lim_{\Delta t \to 0} \frac{\Delta m}{\Delta t}$$

where Δm is the change in mass, Δt is the change in time, and $\lim_{\Delta t \to 0}$ means that Δt becomes very small and almost zero.

Because vectors have a magnitude and direction, the derivative of a vector comprises two terms. The first term represents the change in the magnitude of the vector, and the second term represents the change in the direction of the vector.

1.2.7   Derivative of a unit vector

First, let us start by showing the derivative of the unit vector $\mathbf{e}_r$ (Fig. 1.7), which is pointing in the direction of the vector **r**. The unit vector $\mathbf{e}_r$ can be written in terms of its Cartesian components,

$$\mathbf{e}_r = e_r\cos\theta\mathbf{i} + e_r\sin\theta\mathbf{j}$$

And because the length of a unit vector is 1, that is, $e_r = 1$, then

$$\mathbf{e}_r = \cos\theta\mathbf{i} + \sin\theta\mathbf{j}$$

The derivative of $\mathbf{e}_r$ with respect to the time

$$\frac{d\mathbf{e}_r}{dt} = \frac{d}{dt}(\cos\theta\mathbf{i} + \sin\theta\mathbf{j}) \tag{1.7}$$

Because the derivative is with respect to t and the parameter inside the parentheses is θ, the chain rule will be used. The chain rule is a way to change the derivative with respect to one parameter to a derivative with a different parameter so the derivation can be conducted, such as

$$\frac{dF(y)}{dx} = \frac{dF(y)}{dy} \times \frac{dy}{dx}$$

$$\therefore \frac{d\mathbf{e}_r}{dt} = \frac{d}{d\theta}(\cos\theta\mathbf{i} + \sin\theta\mathbf{j})\frac{d\theta}{dt} = (-\sin\theta\mathbf{i} + \cos\theta\mathbf{j}).\frac{d\theta}{dt}$$

As shown in Fig. 1.7, the vector $(-\sin\theta\mathbf{i} + \cos\theta\mathbf{j})$ represents a unit vector $\mathbf{e}_\theta$, which is perpendicular to $\mathbf{e}_r$.

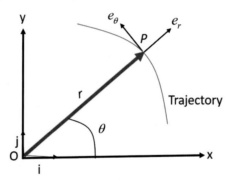

FIGURE 1.7 Representation of a vector in the polar coordinate system.

So the derivative of a unit vector is a unit vector perpendicular to it and multiplied by $(d\theta/dt)$. It will be shown later that $(d\theta/dt)$ is the angular velocity of the vector, that is, how fast the vector is rotating around the center of rotation with respect to a certain reference.

1.2.8 Matrix algebra

A matrix is a rectangular array of rows and columns of numbers enclosed in a bracket (Kreyszig, 1993). Each number in the matrix is called an element. The following is an example of two matrices **A** and **B**, each with two rows, three columns, and six elements.

$$\mathbf{A}_{2\times3} = \begin{bmatrix} 3 & 0.7 & 4 \\ -2 & 1 & 0 \end{bmatrix}; \quad \mathbf{B}_{2\times3} = \begin{bmatrix} 1 & 2 & 0.4 \\ 2.5 & -1 & 3 \end{bmatrix}$$

The algebraic addition and subtraction of these two matrices can be done simply by adding/subtracting the corresponding elements of the two matrices as follows:

$$\mathbf{C}_{2\times 3} = \mathbf{A}_{2\times 3} + \mathbf{B}_{2\times 3} = \begin{bmatrix} 3 & 0.7 & 4 \\ -2 & 1 & 0 \end{bmatrix} + \begin{bmatrix} 1 & 2 & 0.4 \\ 2.5 & -1 & 3 \end{bmatrix} = \begin{bmatrix} 4 & 2.7 & 4.4 \\ 0.5 & 0 & 3 \end{bmatrix}$$

The matrix $\mathbf{C}_{2\times 3}$ can be *transposed* to $\mathbf{C}_{3\times 2}$ by changing its rows with its columns and turning it into a 3×2 matrix as follows:

$$(\mathbf{C}_{2\times 3})^T = \mathbf{C}_{3\times 2} = \begin{bmatrix} 4 & 0.5 \\ 2.7 & 0 \\ 4.4 & 3 \end{bmatrix}$$

Matrix multiplication can be done by multiplying the elements of the rows of one matrix by the elements in the columns of the second matrix. In order to do that, the number of the elements in the rows of the first matrix should equal the number of the elements in the columns of the second matrix. For example, the multiplication of $\mathbf{AB}$ cannot be done because the number of the elements in the rows of the $\mathbf{A}$ matrix is 3 and the number of elements in the columns of the $\mathbf{B}$ matrix is 2. However, the multiplication $\mathbf{A}^T\mathbf{B}$ can be done as follows:

$$\mathbf{D}_{3\times 3} = \mathbf{A}^T_{2\times 3}\mathbf{B}_{2\times 3} = \begin{bmatrix} 3 & -2 \\ 0.7 & 1 \\ 4 & 0 \end{bmatrix} \begin{bmatrix} 1 & 2 & 0.4 \\ 2.5 & -1 & 3 \end{bmatrix} = \begin{bmatrix} -2 & 8 & -4.8 \\ 3.2 & 0.4 & 3.28 \\ 4 & 8 & 1.6 \end{bmatrix}$$

In this process, the element (-2) in the $\mathbf{D}$ matrix, for example, is the result of multiplying the first row of $\mathbf{A}$ by the first column in $\mathbf{B}$ as follows:

$$\begin{bmatrix} 3 & -2 \end{bmatrix} \begin{bmatrix} 1 \\ 2.5 \end{bmatrix} = (3)(1) + (-2)(2.5) = -2$$

The *determinant* of a matrix is an expression associated with a square matrix with an equal number of rows and columns. For example, the determinant of a matrix $\mathbf{E} = \begin{bmatrix} 2 & 3 \\ 1 & 4 \end{bmatrix}$ is as follows:

$$E = det\ \mathbf{E} = det \begin{bmatrix} 2 & 3 \\ 1 & 4 \end{bmatrix} = \begin{vmatrix} 2 & 3 \\ 1 & 4 \end{vmatrix} = 2 \times 4 - 3 \times 1 = 5$$

The determinant of matrices with a higher order, like $\mathbf{D}_{3\times 3}$, can be found by multiplying the elements of each row by a submatrix called minor. The *minor matrix* is a subset of the original matrix and is constructed by deleting the row and column of the element that is associated with the multiplication as follows:

$$D = det\ \mathbf{D} = \begin{vmatrix} -2 & 8 & -4.8 \\ 3.2 & 0.4 & 3.28 \\ 4 & 8 & 1.6 \end{vmatrix} = -2\begin{vmatrix} 0.4 & 3.28 \\ 8 & 1.6 \end{vmatrix} - 8\begin{vmatrix} 3.2 & 3.28 \\ 4 & 1.6 \end{vmatrix} - 4.8\begin{vmatrix} 3.2 & 0.4 \\ 4 & 8 \end{vmatrix} = 0$$

The minus sign before 8 in the second term on the right-hand side is added because the multiplication order takes alternative turns between $+$, $-$, $+$, and so forth.

The *inverse of a matrix* conceptually has the same meaning as the inverse of a number; however, the technique used for inverting a number cannot be generalized to the inverse of a vector or a matrix. While the multiplication of a number by its inverse will result in 1, the multiplication of a matrix by its inverse will result in an identity matrix ($\mathbf{I}$). The *identity matrix* is a matrix with diagonal elements equal to one and zeros for the other elements. For example, the multiplication of a generic matrix, such as $\mathbf{K}$, by its inverse is as follows:

$$\mathbf{K}_{2\times2}\mathbf{K}_{2\times2}^{-1} = \mathbf{I}_{2\times2} = \begin{bmatrix} 1 & 0 \\ 0 & 1 \end{bmatrix}$$

Multiplication of matrix $\mathbf{B}$ and vector $\mathbf{a}$ is similar to the multiplication of two matrices because a vector can be considered a matrix with one row or one column.

$$\mathbf{F}_{1\times3} = \mathbf{a}_{1\times2}\mathbf{B}_{2\times3} = \begin{bmatrix} 1 & 3 \end{bmatrix} \begin{bmatrix} 1 & 2 & 0.4 \\ 2.5 & -1 & 3 \end{bmatrix} = \begin{bmatrix} 8.5 & -1 & 9.4 \end{bmatrix}$$

As can be seen, the result will be a vector. The multiplication of a matrix with a vector is normally seen when solving a set of linear equations where

$$\mathbf{Ax} = \mathbf{b} \tag{1.8}$$

where $\mathbf{A}$ is the matrix with the coefficients of the set of equations, the coefficients of each equation represent one row in $\mathbf{A}$, $\mathbf{x}$ is a vector of unknown parameters, and $\mathbf{b}$ is a vector with known numbers on the right side of each equation.

The solution of the unknown variables in vector $\mathbf{x}$ would be

$$\mathbf{x} = \mathbf{A}^{-1}\mathbf{b} \tag{1.9}$$

1.3   Complex numbers

There are many applications in which complex numbers appear in the calculations (Kreyszig, 1993). Complex numbers emerge when a real solution, such as $x^2 = -1$ or $x = \sqrt{-1}$, exists. Under such circumstances, $x = \sqrt{-1} = i$ is called an imaginary number. By definition, a complex number z has a pair of real numbers x and y and is normally written as $z = x + iy$, where x represents the real part of z and is normally expressed as $x = Re(z)$ and y is the imaginary part and is written as $y = Im(z)$.

Two complex numbers $z_1 = (x_1 + iy_1)$ and $z_2 = (x_2 + iy_2)$ can be added as follows:

$$z_1 + z_2 = (x_1 + iy_1) + (x_2 + iy_2) = (x_1 + x_2) + i(y_1 + y_2) \qquad (1.10)$$

And can be multiplied as follows:

$$z_1 z_2 = (x_1 + iy_1) \times (x_2 + iy_2) = x_1 x_2 + ix_1 y_2 + iy_1 x_2 + i^2 y_1 y_2 = (x_1 x_2 - y_1 y_2) + i(x_1 y_2 + x_2 y_1)$$
$$(1.11)$$

Example:
Let

$$z_1 = (4 + 2i)$$

and

$$z_2 = (6 - 3i)$$

Then

$$\mathrm{Re}(z_1) = 4, \quad \mathrm{Im}(z_1) = 2, \quad \mathrm{Re}(z_2) = 6, \quad \mathrm{Im}(z_2) = -3$$

$$z_1 + z_2 = (4 + 2i) + (6 - 3i) = 10 - i$$

$$z_1 z_2 = (4 + 2i)(6 - 3i) = 30 - 0i = 30$$

$z^* = x - iy$ is called the *conjugate* of $z = x + iy$, where

$$zz^* = (x + iy)(x - iy) = x^2 + y^2 \qquad (1.12)$$

The quotient z_3 of two complex numbers (z_1/z_2) is as follows:

$$z_3 = \frac{z_1}{z_2} = \frac{(x_1 + iy_1)}{(x_2 + iy_2)} \qquad (1.13)$$

Multiplying the numerator and denominator by the conjugate of the denominator,

$$z_3 = \frac{z_1}{z_2} = \frac{z_1 z_2^*}{z_2 z_2^*} = \frac{(x_1 + iy_1)(x_2 - iy_2)}{(x_2 + iy_2)(x_2 - iy_2)} = \frac{x_1 x_2 + y_1 y_2}{x_2^2 + y_2^2} + i\frac{x_2 y_1 - x_1 y_2}{x_2^2 + y_2^2} = x_3 + iy_3$$

where $x_3 = ((x_1 x_2 + y_1 y_2)/(x_2^2 + y_2^2))$ is the real part of z_3 and $y_3 = ((x_2 y_1 - x_1 y_2)/(x_2^2 + y_2^2))$ is the imaginary part of z_3.

The complex numbers can also be presented in the polar coordinate system in terms of r and θ. In this case, $x = r\cos\theta$ and $y = r\sin\theta$, and then $z = x + iy$ takes the polar form $z = r(\cos\theta + i\sin\theta)$.

The *absolute value* or the *modulus* of z, which is denoted by $|z|$, can be expressed as follows:

$$|z| = r = \sqrt{x^2 + y^2} = \sqrt{zz^*} \qquad (1.14)$$

The angle θ is called the *argument* of z or the phase angle and is calculated as follows:

$$\theta = \arg(z) = \tan^{-1}\left(\frac{y}{x}\right) \tag{1.15}$$

Example:
Let $z = 3 + 4i$ in the Cartesian coordinate system, then $r = \sqrt{3^2 + 4^2} = 5$, $\theta = \tan^{-1}\left(\frac{4}{3}\right) = 53.13°$, $|z| = r = 5$, and $z = 5(\cos(53.13) + i\sin(53.13))$ in the polar coordinate system.

The multiplication of two complex numbers $z_1 = r_1(\cos\theta_1 + i\sin\theta_1)$ and $z_2 = r_2(\cos\theta_2 + i\sin\theta_2)$ in the polar form can be expressed as follows:

$$z_1 z_2 = r_1 r_2[\cos(\theta_1 + \theta_2) + i\sin(\theta_1 + \theta_2)]$$

Example:
Let

$$z_1 = (6 - 3i)$$

and

$$z_2 = (4 + 2i)$$

for

$$z_1 \Rightarrow r_1 = \sqrt{6^2 + (-3)^2} = \sqrt{45} = 6.71; \quad \theta_1 = \tan^{-1}\left(\frac{-3}{6}\right) = -26°.56$$

and

$$z_2 \Rightarrow r_2 = \sqrt{4^2 + (2)^2} = \sqrt{20} = 4.47; \theta_2 = \tan^{-1}\left(\frac{2}{4}\right) = 26.56°$$

$$z_1 = 6.71(\cos(-26.56) + i\sin(-26.56)); \quad z_2 = 4.47(\cos(-26.56) + i\sin(-26.56))$$

$$z_1 z_2 = r_1 r_2[\cos(\theta_1 + \theta_2) + i\sin(\theta_1 + \theta_2)] = (6.71)(4.47)[\cos(-26.56 + 26.56) + i\sin(-26.56 + 26.56)]$$

$$z_1 z_2 = 30[\cos(0) + i\sin(0)] = 30$$

1.3.1 Hermitian matrix

For some applications, the elements of the matrix contain complex numbers, and the matrix becomes a *complex matrix*. A *Hermitian matrix* is a square complex matrix that satisfies the following condition:

If $\mathbf{A} = \begin{bmatrix} 5 + 2i & -4 \\ 8i & 9 - 3i \end{bmatrix}$ is a square complex matrix, then $\overline{\mathbf{A}}^T = \begin{bmatrix} 5 - 2i & -8i \\ -4 & 9 + 3i \end{bmatrix}$ is a Hermitian matrix that represents the transpose of matrix A and has elements that are the conjugates of the elements of A.

1.3.2   Unitary matrix

A *unitary matrix* **U** is a complex square matrix whose conjugate transpose **U*** is equal to its inverse. Mathematically:

$$\mathbf{U}^*\mathbf{U} = \mathbf{U}\mathbf{U}^* = \mathbf{I}$$

1.3.3   Numerical differentiation of time signals

The time signals resulting from a system's response to vibration are normally collected using data acquisition systems and are normally presented in discrete forms. In this case, different types of data processing operations can be done, including differentiation and integration, numerically using state-of-the-art numerical techniques. The *finite difference method* is a very popular data differentiation method. In this process, the function, which could be in the time or spatial domain, is discretized using a *Taylor series expansion*.

$$f'(x) = \lim_{h \to 0} \frac{f(x+h) - f(x)}{h} \tag{1.16}$$

where $f'(x)$ is the derivative of the function ($f(x)$) at x, and h is the distance between adjacent data points. There are different forms of the finite difference method, including the *forward*, *backward*, and *central differences*. Eq. (1.16) represents the forward difference approach, where the function value at the current state is subtracted from the function value at a future location $x + h$ and the difference is then divided by the h.

In the backward difference method, the previous function value at a location $x - h$ is subtracted from the current function value at x and then divided by h as follows:

$$f'(x) = \frac{f(x) - f(x-h)}{h} \tag{1.17}$$

If the function values are available in the past and in the future, then the central difference method is a good approach to find the numerical derivative of the function. In this case,

$$f'(x) = \frac{f(x+h) - f(x-h)}{2h} \tag{1.18}$$

Higher-order derivatives can be achieved with the finite difference method using more terms from the Taylor series expansion; however, one should be careful when using the finite difference method with noisy data and data of nonlinear behaviors. In this case, the finite difference method would be very sensitive to the size of h selected in the equation and could produce erroneous results.

1.3.4 Motion of a particle

The motion of any object generally takes place and changes direction in the three-directional spatial space, the space we are living in; therefore it is described using the concepts of vectors. The motion of a point can be described through its displacement, velocity, and acceleration. Fig. 1.8 shows the simplest type of motion represented by the motion of a point on a straight line; this could be a person walking on a straight road between points A and B. In order to specify the position of a point in space, a coordinate reference system must be established. A Cartesian coordinate system (x, y, and z) (Fig. 1.8) is the most traditionally used system. The Cartesian coordinate system has three orthogonal, or perpendicular, axes, representing the space surrounding us. The point at which the three axes intersect is called the origin (O).

The origin represents the reference point in the coordinate system with location ($x = 0$, $y = 0$, and $z = 0$). Once the coordinate system is established, the location of the point relative to this origin can be easily monitored. The reference system can be fixed in space or to the ground, such as a reference system in a room that can be fixed to one of its corners; in this case, the positions of people sitting in this room can be identified with respect to the origin of the reference system. At the same time, the relative position between the people sitting and moving in the room can also be identified. Furthermore, the reference system can be attached to moving objects, such as a bus or an airplane. In this case, the reference system can be rigidly attached to the object and can move and rotate with it. Now if a person is standing in a bus station, they will see the bus moving with the people inside it. In this case, we call the reference system

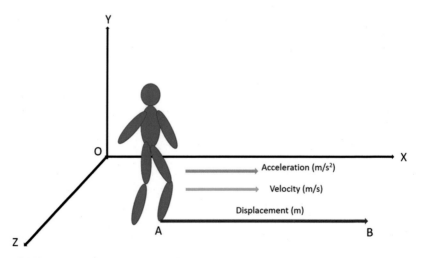

FIGURE 1.8 Straight-line motion between point A and point B where the displacement, velocity, and acceleration are all pointing in the same direction.

that is fixed to the ground at the bus station a *global system*, and we call the reference system that is attached to the bus and moving with the bus a *body-fixed* or *secondary coordinate system*. The people sitting in the bus will have a different experience; they see all people in the bus as not moving, but they see the person in the bus station moving relative to them. Care should be taken when using local reference systems as they can generate errors in the motion calculation if their acceleration and rotation are not taken into account.

1.3.5   Displacement, velocity, and acceleration

The displacement of a point is the distance the point travels with time. The displacement can be on a straight line, such as a person walking on a straight walkway, or it can be on a curved or circular path, such as a car merging onto a highway.

Velocity **v** is the rate of change of displacement **s** with time.

$$\mathbf{v}(t) = \frac{d\mathbf{s}}{dt} \tag{1.19}$$

Because it is the derivative of the displacement, the velocity is generally a tangent (like a tangent to a circle) to the displacement path. The magnitude of velocity is called speed.

The derivative of the velocity **v** with respect to time is called acceleration **a**.

$$\mathbf{a}(t) = \frac{d\mathbf{v}}{dt} \tag{1.20}$$

Acceleration normally has two terms. One term represents the change of the magnitude of the velocity with time, and the second term represents the change in the velocity direction with time.

Acceleration is normally a function of time and can be measured much more easily than velocity and displacement using a device called an *accelerometer*. In such a case, the velocity can be calculated by integrating the acceleration with time, and then the displacement can be determined by integrating the velocity with time. The calculation of velocity from acceleration can be done as follows:

$$\mathbf{a}(t) = \frac{d\mathbf{v}}{dt} \Rightarrow d\mathbf{v} = \mathbf{a}(t)dt \Rightarrow \int_{v_0}^{v} d\mathbf{v} = \int_{t_0}^{t} \mathbf{a}(t)dt \Rightarrow \mathbf{v} = \mathbf{v}_0 + \int_{t_0}^{t} \mathbf{a}(t)dt \tag{1.21}$$

where $\mathbf{v}_0$ and t_0 are the initial velocity and initial time, respectively.

For applications in hydrodynamics and aerodynamics, the acceleration can be a function of the velocity. In such cases, the separation of variables is used, and the velocity can be calculated as

$$a(v) = \frac{dv}{dt} \Rightarrow dt = \frac{dv}{a(v)} \Rightarrow \int_{t_0}^{t} dt = \int_{v_0}^{v} \frac{dv}{a(v)} \Rightarrow t = t_0 + \int_{v_0}^{v} \frac{dv}{a(v)} \tag{1.22}$$

For situations such as those in gravitational applications and springs analysis, the acceleration can be a function of the displacement. In this case, the chain rule and separation of variables must be used, and the velocity can be calculated as follows:

$$a(s) = \frac{dv}{dt} = \frac{dv}{ds}\frac{ds}{dt} = \frac{dv}{ds}v \Rightarrow a(s)ds = vdv \Rightarrow \int_{s_0}^{s} a(s)ds = \int_{v_0}^{v} vdv \Rightarrow v = v_0 + \int_{s_0}^{s} a(s)ds$$

$$(1.23)$$

where s_0 is the initial displacement.

1.3.6 Curvilinear motion

When a point moves in space, it changes its position with time; also, it can change its direction. One example is driving a car on a curved road. In this case, when the driver increases the speed of the car, it changes the magnitude of the motion; during this time, the car is changing its direction as it is moving on the curved road. The motion in this case can be expressed in terms of tangential and normal components or by using a polar coordinate system.

1.3.6.1 Tangential and normal components

In a tangential and normal system, the velocity **v** will be tangent to the path of motion (Fig. 1.9) and will represent the change in the displacement (s) on the curved path with respect to time. Because the velocity can change in magnitude and direction, the acceleration, which is (dv/dt), will have two components. One is tangential in the direction of the velocity and will be proportional to the change in the velocity magnitude $a_t = (dv/dt)$. The second one is perpendicular to it and pointing to the center of the path, which is traditionally called normal, or *centripetal*, acceleration. Normal acceleration is a result of the change in the direction of the velocity, meaning that this component will be zero when moving in a straight line. The normal component is proportional to $a_n = (v^2/\rho)$, where ρ is the instantaneous radius of the curved path and v is the magnitude of the velocity at the specified ρ. The magnitude of the resultant acceleration will be $a = \sqrt{a_t^2 + a_n^2}$.

For angular motion where the radius of curvature is constant, $\rho = R$, that is, circular motion (Fig. 1.10), the expression for the velocity becomes $v = \omega R$, $a_t = \alpha R$ for the tangential acceleration and $a_n = \omega^2 R$ for normal or centripetal acceleration. In this case, the angle θ represents the angular displacement, $\omega = d\theta/dt$ is the angular velocity, and $\alpha = d\omega/dt$ is the angular acceleration.

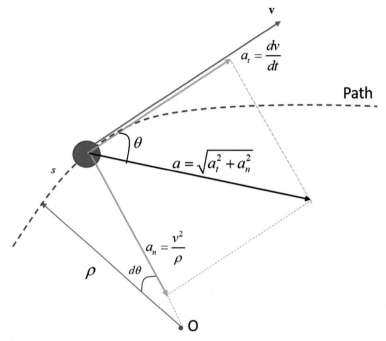

FIGURE 1.9   Tangential and normal representation of the motion of a particle on a curvilinear path.

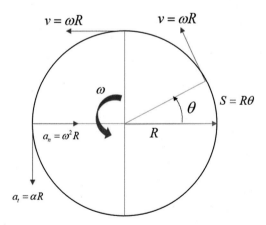

FIGURE 1.10   Motion of a particle on a circular path.

Example:

A car is moving on a circular path with a radius of $R = 50$ m at angular velocity $\omega = 5$ rad/s. Determine the car's velocity and acceleration.

Solution:

$$\omega = 5 \ \text{rad}/s = \text{constant}, \quad \text{therefore} \ \alpha = \frac{d\omega}{dt} = 0$$

$$v = \omega R = (5)(50) = 250 \ m/s$$

$$a_t = \alpha R = (0)(50) = 0$$

$$a_n = \omega^2 R = (5)^2(50) = 1250 \ m/s^2$$

1.3.6.2 Polar components

If the vector $\mathbf{r} = r\mathbf{e}$ is expressed as a general vector in the polar coordinate system (Fig. 1.11), then it can change in magnitude and direction with time, where r is the magnitude of the vector and $\mathbf{e}$ is a unit vector in the direction of the vector. The derivative of $\mathbf{r}$ with time represents the velocity $\mathbf{v}$.

In this case, the velocity v has two components representing the change in magnitude v_r and the change in direction v_θ of the vector r as follows:

$$\mathbf{v} = v_r\mathbf{e}_r + v_\theta\mathbf{e}_\theta$$

where $v_r = dr/dt$ in the direction of $\mathbf{r}$ and $v_\theta = r(d\theta/dt)$ in the direction perpendicular to $\mathbf{r}$, which is traditionally called the θ-direction as it is pointing in the direction of θ.

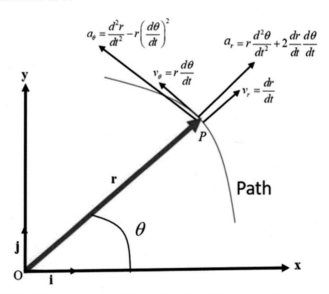

FIGURE 1.11 Polar representation of the motion of a particle on a curvilinear path.

The acceleration will also have two components, given as follows:
$\mathbf{a} = a_r\mathbf{e}_r + a_\theta\mathbf{e}_\theta$, where $a_r = r(d^2\theta/dt^2) + 2(dr/dt)(d\theta/dt)$ in the $\mathbf{r}$ direction, and $a_\theta = (d^2r/dt^2) - r(d\theta/dt)^2$ in the θ-direction. The angle θ is called

the *angular displacement*, measured in radians (rad), and it represents the angular distance that a line, such as **r**, can make relative to a reference line, such as *x* (Fig. 1.11). The derivative $d\theta/dt$ is the angular velocity ω in rad/s, and the second derivative of θ, $((d^2\theta/dt^2) = (d\omega/dt))$ is the angular acceleration α and is measured in rad/s^2.

1.4 Forces and motion

Kinematics is the study of the motion of a particle or any object with no consideration of forces associated with it. On the other hand, *kinetics* is the study of motion that takes into consideration the forces that are accompanying the motion and interacting with the object. Newton's second law presents a mathematical representation of the relationship between motion and force. It states that the summation of external forces is equal to the rate of change of the linear momentum (the multiplication of the mass by the velocity) of a particle, which can be written as follows:

$$\sum \mathbf{F} = \frac{d}{dt}(m\mathbf{v}) \tag{1.27}$$

where $\sum \mathbf{F}$ is the summation of external forces applied on the object, including gravity forces (the weight of the object), and is a vector quantity; m is the mass of the object and is a scalar quantity; and **v** is the velocity of the particle and is a vector quantity. Assuming the mass is rigid and not changing in magnitude with time, the relationship between the motion of the particle and the forces applied on it is

$$\sum \mathbf{F} = \frac{d}{dt}(m\mathbf{v}) = m\frac{dv}{dt} = m\mathbf{a} \tag{1.28}$$

where **a** is the acceleration of the particle, which is a vector. In SI units, the unit for force is Newton (N), the unit for mass is kilograms (kg), and the unit for acceleration is m/s^2; in United States customary units, the units are pound (lb), slug, and foot/s^2, respectively.

For a motion on a circular path, Newton's second law, or the equation of motion, can be expressed as follows:

$$\sum \mathbf{F} = \sum F_t\mathbf{e}_t + \sum F_n\mathbf{e}_n = m(a_t\mathbf{e}_t + a_n\mathbf{e}_n) \tag{1.29}$$

where

$$\sum F_t = ma_t = m\frac{dv}{dt}$$

is the equation of motion in the tangential direction to the path of motion and

$$\sum F_n = ma_n = m\frac{v^2}{\rho}$$

is the equation of motion in the centripetal direction to the path of motion (Fig. 1.12).

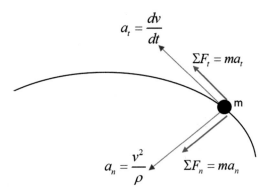

FIGURE 1.12   Tangential and normal representation of forces and motions of a particle moving on a curvilinear path.

Example:

A car with a mass of $m = 1000$ kg is moving at a constant speed of $v = 12\ m/s$ on a curved road. The radius of curvature of the road is $\rho = 100$ m. Find the forces on the car in the tangential and perpendicular directions to the curve. Neglect frictional forces.

$\sum F_t = ma_t = m\frac{dv}{dt} = 0$ because v is constant.

$$\sum F_n = ma_n = m\frac{v^2}{\rho} = 1000\frac{(12^2)}{100} = 1440N$$

1.4.1   Kinetics of rigid bodies

For rigid bodies, Newton's second law can be applied with the consideration that the acceleration is the acceleration of the center of mass of the rigid body. The difference between a rigid body and a particle is that the rigid body has dimensions, so it can translate like a particle and can also rotate. In this case, the equation of motion of a rigid body is conceptually equivalent to two motions: translation and rotation. The translational motion follows the following equation:

$$\sum \mathbf{F} = m\mathbf{a} \qquad (1.30)$$

where $\sum \mathbf{F}$ is the summation of the external forces applied on the body, including its weight, and $\mathbf{a}$ is the acceleration of the center of mass of the rigid body.

The rotational motion can be expressed as follows:

$$\sum M = I\alpha \tag{1.31}$$

where $\sum M$ is the summation of the external applied moments (a moment represents a force multiplied by its perpendicular distance of action from a certain reference point) about the center of mass of the object and I is the moment of inertia of the rigid body. The moment of inertia is a mathematical form that indicates how the mass of the rigid body is distributed away from its center. Large moments of inertia indicate a larger distribution of materials away from the center of mass and present higher resistance to rotation. Formulas for I of objects with different shapes can be found in books such as *Engineering Dynamics* (Bedford & Fowler, 2008); α is the angular acceleration of the rigid body about its center of mass.

Example:

A disk is rolling on the ground as shown in Fig. 1.13. The mass of the disk is 20 kg, its radius R is 200 mm, and it is subjected to an applied moment of 50 N m. Find the disk's angular acceleration α and the acceleration of its center of mass at O.

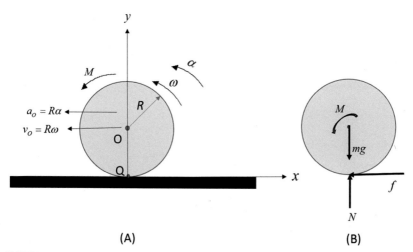

(A)   (B)

FIGURE 1.13   (A) Motion of a rolling disk, (B) free-body diagram showing the external forces applied to the disk including the applied moment (M), normal ground reaction force (N), weight (mg) comprising the mass multiplied by gravity, and friction force (f).

Solution:

In dynamics analysis, the concept of a free-body diagram presents a critical first step in the solution process. The free-body diagram shows all external forces, including the reaction forces, applied on the body and their directions (Fig. 1.13B). In this case, the disk will be subjected to a normal ground reaction force (N) as a reaction to the disk weight (mg) and the friction force (f) that opposes slippage with respect to the floor that can happen because of the applied moment (M).

In rolling motion, without slippage, the velocity of point Q on the disk is considered to be zero because, at the instant shown, point Q can also be considered as a point on a nonmoving ground with zero velocity. So, at this instant, Q can be considered as a fixed point that the whole disk is rotating about. Using the expression for circular motion, the velocity of the center of the disk O with respect to Q can be calculated as $v_O = R\omega$. The angular velocity ω and the angular acceleration α are traditionally considered positive if they are pointed in the counterclockwise direction.

The acceleration of the center of the disk at O is the derivative of the velocity of the center of the disk with respect to time, that is,

$$a_O = \frac{dv_O}{dt} = \frac{d(R\omega)}{dt} = R\frac{d\omega}{dt} = R\alpha \Rightarrow \alpha = \frac{a_O}{R} \tag{1.32}$$

Applying the equation of motion in the x-direction,

$$\Sigma F_x = ma_x \Rightarrow -f = -ma_O \tag{1.33}$$

Applying the moment equation about the center of the disk,

$$\Sigma M_O = I_O\alpha \Rightarrow M - fR = I_O\alpha \tag{1.34}$$

The moment of inertia of a disk about its center of mass is

$$I_O = \frac{1}{2}mR^2$$

Changing the unit of R from millimeters to meters:

$$R = \frac{200}{1000} = 0.2 \text{ m}$$

$$\therefore I_O = \frac{1}{2}(20)(0.2)^2 = 0.4 \text{ kg m}^2$$

Substitute (1.32) and (1.33) in (1.34):

$$50 - (ma_O)R = I_O\frac{a_O}{R} \Rightarrow 50 - (20a_O)(0.2) = 0.4\frac{a_O}{0.2} \Rightarrow 50 - 4a_O = 2a_O \Rightarrow a_O = \frac{50}{6} = 8.33 \ m/s^2$$

Substitute in (1.32),

$$\alpha = \frac{a_0}{R} = \frac{8.33}{0.2} = 41.65$$

rad/s^2 in the counterclockwise direction.

1.4.2 Forces and motion of human-body segments

In dynamics analysis of the human body, the human body can be simplified as a collection of concentrated masses connected via springs and dampers. For example, the head can be considered a rigid sphere connected to the neck (another rigid segment), and so on for the rest of the body segments. The spring and dampers, respectively, represent the resistance to deformation and the ability of the muscles and surrounding tissues to absorb and dissipate energy. In this case, the equations of motion can be applied to each segment of the human body to investigate the amount of force it is experiencing because of motion. Alternatively, motions resulting from the application of forces can also be quantified.

The motion of the segments of the human body can be very complicated. Take the head and neck for example; even if we assume the neck is one piece or a rigid link, the motion of the head itself can have many components. The head can rotate with pitching motion, similar to nodding yes; yaw or twisting motion, like shaking the head no; and lateral motion, like tilting the head toward the shoulder. In addition to these rotational motions, because the head is connected to the neck via a flexible joint representing the neck tissues and vertebrae, there is a tendency for the head to move away from the neck and force the joint to stretch, compress, and shear. Now, if the neck is considered flexible and comprises several components, the analysis will become even more complicated, as we have to deal with the motion of flexible multibody dynamics. Fortunately, commercial programs like Abaqus and advanced imaging techniques are currently available and are normally used by researchers to create complex models of human segments and provide the capability to test them under different types of loading.

Example:

In this example (Liu, 2011; Rahmatalla & Liu, 2012), the human head−neck is modeled with a planar simple mechanical system that comprises a solid circular mass that represents the head and a rigid link that represents the neck (Fig. 1.14). The head and neck are rigidly attached in this model, acting like an inverted pendulum. The head−neck model is connected to the seventh cervical (C7) vertebra via a hinge joint, so it can rotate around it with angular displacement q. Other elements can be added to the head−neck model to make it more realistic, including a spring with a stiffness k that represents the resistance of the cervical spine and surrounding tissues to deformation and a damper c that reflects the capability of the tissues to absorb and dissipate the energy and external motions entering the body. An active muscle model τ can also be added to the system to represent the active forces that the muscles induce to resist external motions and to balance and hold the head in a certain posture. Of course, models with more detail can be developed using advanced commercial software packages, but a model like this would be used to give a general idea about the vibration induced on the human head−neck area of drivers or pilots, especially when seat belts and straps are used to hold their bodies against their seats or supporting surfaces.

In addition to the motion of each body segment, the relative motion between adjacent segments of the human body should be studied to provide more insight

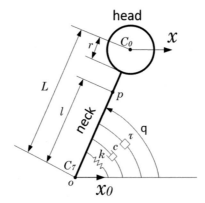

FIGURE 1.14   Schematic representation of a planar human head−neck model. *Adapted from Rahmatalla, S., & Liu, Y. (2012). An active head−neck model in whole-body vibration: Vibration magnitude and softening.* Journal of Biomechanics, 45(6), 925−930.

into what is really happening inside the human body at certain locations. For example, the relative motion between the head and torso would provide important information about what is happening in the neck area. This is important because the torso has a much bigger mass than the head−neck region. During prehospital transport, the supine human is normally immobilized so that head motion is restricted to protect the head−neck area from secondary damage. If the driver of the ambulance brakes frequently, for example, the sliding motion of the heavy torso relative to the relatively fixed head can generate a large force on the neck area that can be reflected as tension and compression. So even if the person is wearing a neck collar, it will not protect the person from the consequences of tension/compression motions that can cause discomfort, pain, and potential secondary injuries under extreme conditions.

1.5   Basic statistics

This section presents the definition of statistics metrics used in this book.

1.5.1   Mean or average

The average, or mean, is an estimate of the value most representative of the population. The mean represents the sum of all data in the data set divided by the number of data points. The calculation of the mean of a data set can be accomplished using the following equation:

$$\text{Mean} = \frac{(x_1 + x_2 + x_3 + \ldots + x_N)}{N} \tag{1.35}$$

where x is a data point and N is the number of the data points in the data set.

For example, the mean for the data set [2, 5, 7, 9, 3, 10, 13] is:

$$\text{Mean} = \frac{(2 + 5 + 7 + 9 + 3 + 10 + 13)}{7} = 7$$

1.5.2 Median

The median represents the value in the middle of the list between the lower half and upper half of the data set. For example, the median for the data set [2, 5, 7, 9, 3, 10, 13] is 7. If the number of data points is even, then the median is calculated by taking the average of the two values in the center of the data. For the data set [2, 5, 7, 9, 3, 10], for example, the median can be calculated by rearranging the data set to [2, 3, 5, 7, 9, 10]; the median is $(5 + 7)/2 = 6$.

The median can be a better measure or representative of the data set when there are outliers (points far away from other points in the set). For example, for the data set [2, 3, 5, 7, 9, 10, 13, 100], the mean will be 18.6; the median $(9 + 7)/2 = 8$ will be less affected by the outlier (100) and becomes 8 instead of 7.

1.5.3 Range, variance, and standard deviation

The range represents the difference between the magnitude of the lowest and highest values in the data set. The range is a good measure of the variability in the data set.

The variance gives an idea about how much the data is spread from the mean. The variance is calculated by taking the difference between each data point in the data set and the mean of the data set, squaring that difference, and repeating that for all data points in the data set. The variance is calculated by summing all square differences and then dividing the sum by the number of points in the data set minus 1, as follows:

$$\text{Variance} = \frac{((x_1 - x')^2 + (x_2 - x')^2 + (x_3 - x')^2 + \ldots + (x_N - x')^2)}{N - 1} \quad (1.36)$$

where x' is the mean.

The variance for the data set [2, 5, 7, 9, 3, 10, 13] is

$$\text{Variance} = \frac{((2-7)^2 + (5-7)^2 + (7-7)^2 + (9-7)^2 + (3-7)^2 + (10-7)^2 + (13-7)^2)}{7 - 1} = 15.67$$

The standard deviation (σ) is the square root of the variance and can be mathematically represented as follows:

$$\text{Standard deviation } (\sigma) = \sqrt{\text{Variance}}$$

In this case, $\sigma = \sqrt{15.67} = 3.96$.

1.5.4   Root mean square

The root mean square (RMS) is a single number that represents the magnitude of the signal. RMS is a very popular measure in vibration analysis. Its mathematical calculation is relatively similar to that of standard deviation, given as follows:

$$\text{RMS} = \sqrt{\frac{\left((x_1)^2 + (x_2)^2 + (x_3)^2 + \ldots + (x_N)^2\right)}{N}} \tag{1.37}$$

The RMS for the data set [2, 5, 7, 9, 3, 10, 13] is

$$\text{RMS} = \sqrt{\frac{\left((2)^2 + (3)^2 + (5)^2 + (7)^2 + (9)^2 + (10)^2 + (13)^2\right)}{7}} = 7.9$$

1.5.5   Regression analysis: least-squares line fitting

Least-squares line fitting is a methodology used to find the best line that can pass through a data set. This can be obtained when the sum of the squares of the residuals (the difference between the location of a data point and the corresponding point on the line) is as small as possible (Fig. 1.15). The residuals (R_i) represent the vertical deviations of the data points from a proposed line with a mathematical form $Y = mX + b$.

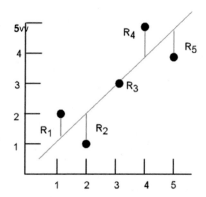

FIGURE 1.15   Best line fit between data point and the residuals (R_i).

The goal in the regression process is to minimize the vertical distance R_i between the points and the line passing between the points. The parameters m and b can be calculated from the following expressions (Eide et al., 2018):

$$m = \frac{n\left(\sum x_i y_i\right) - \left(\sum x_i\right)\left(\sum y_i\right)}{n\sum (x_i)^2 - \left(\sum x_i\right)^2} \tag{1.38}$$

$$b = \frac{\sum y_i - m\left(\sum x_i\right)}{n} \tag{1.39}$$

After finding the coefficients m and b, the next step is to test the goodness of the fit (the linear relationship) between Y and X using the least-squares method. The first test is to use the correlation coefficient (r), which has a value between -1 and 1, that is, $-1 \le r \le 1$. $r = 1$ means it is perfectly correlated with the positive slope, where all points are close to the line fit and any value of x can give a very good prediction of y; $r = -1$ means it is perfectly correlated with the negative slope; and $r = 0$ means there is no correlation between variables and the points are scattered about a poor fit. The mathematical expression for calculation r is as follows:

$$r = \frac{n\left(\sum x_i y_i\right) - \left(\sum x_i\right)\left(\sum y_i\right)}{\sqrt{n\sum(x_i^2) - \left(\sum x_i\right)^2}\sqrt{n\sum(y_i^2) - \left(\sum y_i\right)^2}} \tag{1.40}$$

The second indicator of the quality of the line fitting between X and Y is the coefficient of determination (R^2), which is the square of the correlation coefficient (r). The coefficient of determination can have a value between 0 and 1, that is, $0 \le R^2 \le 1$, with $R^2 = 1$ indicating a perfect fit.

Example:

Find the best fit for the data $x = [3, 1, 5, 7, 4]$ and $y = [6, 3, 7, 10, 8]$ using the method of least squares.

The first step is to draw a table containing the values of x and y and the components shown in the m and b expressions as follows:

	x	y	xy	x^2	y^2
	3	6	18	9	36
	1	3	3	1	9
	5	7	35	25	49
	7	10	70	49	100
	4	8	32	16	64
Sum $\sum$	20	34	158	100	258

Substitute in the m and b from Eqs. (1.38) and (1.39):

$$m = \frac{n\left(\sum x_i y_i\right) - \left(\sum x_i\right)\left(\sum y_i\right)}{n\sum (x_i)^2 - \left(\sum x_i\right)^2} = \frac{5(158) - (20)(34)}{5(100) - (20)^2} = 1.1$$

$$b = \frac{\sum y_i - m\left(\sum x_i\right)}{n} = \frac{34 - 1.1(20)}{5} = 2.4$$

The equation for the best line fit for this set of data is $Y = 1.1X + 2.4$, as shown in Fig. 1.16.

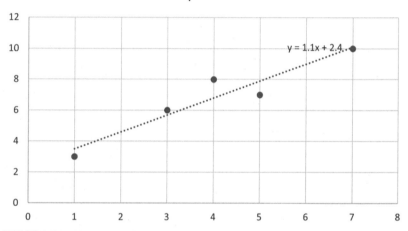

FIGURE 1.16   The best line fit using the least-squares method in Microsoft Excel.

The correlation coefficient (r) from Eq. (1.40) is

$$r = \frac{n\left(\sum x_i y_i\right) - \left(\sum x_i\right)\left(\sum y_i\right)}{\sqrt{n\sum (x_i^2) - \left(\sum x_i\right)^2}\sqrt{n\sum (y_i^2) - \left(\sum y_i\right)^2}} = \frac{5(158) - (20)(34)}{\sqrt{5(100) - (20)^2}\sqrt{5(258) - (34)^2}} = 0.95$$

and the coefficient of determination (R^2) is

$$R^2 = (0.95)^2 = 0.903$$

1.5.6   Distribution of data

The distribution of natural data can be illustrated by a curve representing the normal distribution of the data that follows certain rules. As shown in Fig. 1.17, 68% of the data is within one standard deviation from the mean, 95% of the data falls within two standard deviations from the mean, and 99.7% of the data falls within three standard deviations from the mean.

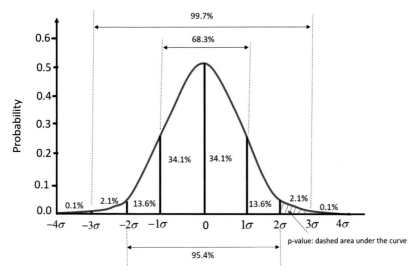

FIGURE 1.17 Normal distribution representation of data point showing with a mean at zero and the probability distribution with 68.3%, 95.4%, and 99.7%.

1.5.7 Confidence intervals

Confidence intervals provide information about the range in which the true value or the mean lies within a certain degree of probability. The size of the confidence interval depends on the sample size and the standard deviation. A narrower confidence interval is normally associated with a large sample size, and a wider confidence interval indicates a small sample size. If the data is highly dispersed, the conclusion becomes less certain, and the confidence interval becomes wider. A 95% confidence interval (CI) means that there is 95% confidence that the sample mean or the true result lies in this interval, and a 99% confidence interval is wider than a 95% confidence interval. The 95% confidence interval can be calculated as follows:

$$CI = \text{sample mean} \pm 1.96 \times \sigma \qquad (1.41)$$

1.5.8 Probability value

The probability value, or p-value, indicates the measure of evidence against the hypothesis. Small p-values correspond to strong evidence. The p-value represents the shaded area under the normal distribution graph in Fig. 1.17. If the p-value is below a predefined limit (typically $p < .05$ in medical research), it is defined as statistically significant.

The p-value is calculated from statistical tables based on the values of the Z-score. The Z-score represents any score value from a normal distribution, such as that shown in Fig. 1.17, with a mean value of μ and a standard deviation of σ, that can be transformed to the normal distribution using the following formula:

$$Z = \frac{x - \mu}{\sigma} \qquad (1.42)$$

The Z-score is positive if the value lies above the mean and negative if it lies below the mean.

1.6   Time and frequency domain analysis

Motion is normally described and measured in the time domain using specialized sensors. This means that the displacement of a point, for example, can be correctly identified at any time during its motion history. In vibration, however, the motion is traditionally transformed using spectrum analysis software from the time domain to the frequency domain, where many analyses can be more easily conducted. The transformation of the signal from the time domain to the frequency domain will distribute the energy of the signal into its components across the frequency spectrum. For example, if the shape of the motion in the time domain is like a sine wave with 50 cycles per second (Fig. 1.18A), that is, 50 Hz, then the transformation of this signal to the frequency domain will look like a vertical line crossing the frequency at 50 Hz (Fig. 1.18B). The signal will look like one line in the frequency domain where all the energy is projected at that frequency. In the frequency domain, the energy of a signal at each frequency can be plotted.

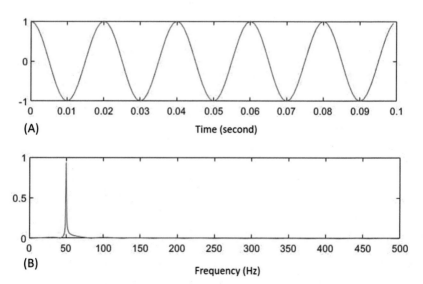

FIGURE 1.18   Time and frequency domain of a sine signal moving at 50 Hz: (A) time domain, (B) frequency domain.

1.6.1   Fourier transform

Fourier transform is the process of transforming a periodic signal from the time domain to the frequency domain using a series of sine and cosine functions. Fourier transform can be also applied to nonperiodic signals using special techniques. The Fourier transform of a signal into its components with their corresponding energies is analogous to the transformation of light to its color components and spectral representations and density. While the calculation of the Fourier transform can be done theoretically (Kreyszig, 1993), existing commercial software such as MATLAB can be used for very complex functions. The Fast Fourier Transform FFT(x) function in MATLAB can be used to transform the function from the time domain to the frequency domain.

The signal in the time domain represents the signal values at different times. While many useful analyses and metrics can be defined in the time domain, including the calculation of the RMS, analysis in the frequency domain is more popular in vibration analysis. Fig. 1.19 shows examples of a sine function with different frequency components. As shown in Fig. 1.19A, the function (sin x) is composed of two frequencies of 50 and 120 Hz in the time domain. Fig. 1.19B depicts the latter signal in the frequency domain. In this case, the graph shows two peaks, one at 50 Hz and the second at 120 Hz. It is evident that more energy is associated with 120 Hz than 50 Hz as the line representing 120 Hz is higher than the one for 50 Hz.

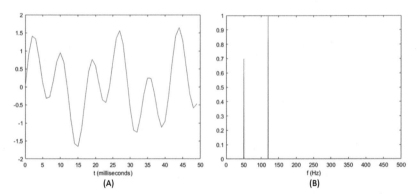

FIGURE 1.19   Time and frequency domain signals: (A) sine signal that has 50 and 120 Hz in the time domain, (B) sine signal that has 50 and 120 Hz in the frequency domain.

1.7   Vibration fundamentals

The motion of an object during vibration occurs when the object is exposed to a cyclic motion such as going up and down while passing through the equilibrium position or a reference position, and it can happen in a regular or random manner. The equilibrium position is the location of the object when it is in a static position and before the vibration is applied. Fig. 1.20 shows an example of the vibratory oscillatory motion of a pendulum. In this case, position O represents the equilibrium position. When the pendulum is put in position A, for example, it will oscillate and pass by position O to position B and then swing back to A. This oscillatory motion can go for an infinite time if there is no friction.

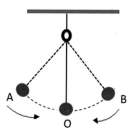

FIGURE 1.20 Oscillatory motion of a pendulum.

Vibration motion can be classified as free, forced, damped, and undamped. In free vibration, the system is freely vibrating with no external forces acting on it. One example is pulling a mass attached to a pendulum (Fig. 1.20) and then letting it go to vibrate freely. If the mass is vibrating inside a liquid, for example, the friction between the mass and the liquid will generate damping that is proportional to the viscosity of the liquid; under such circumstances, the vibration is called free-damped vibration. If a force is applied to the system, such as placing the pendulum on a shaking table, the vibration coming from the shaker will force the pendulum to continuously vibrate; in this case, the vibration is called forced vibration.

Vibration can also be classified as harmonic, periodic, and nonperiodic (Meirovitch, 1975). Examples of harmonic motion are the sine and cosine functions that repeat their motion during each cycle. This could be similar to sitting on a swing with a repeated motion. The cycle of a signal in vibration is one complete motion of the object while it is going up, going down, and then returning to its starting position. The time spent in each cycle is called a period. Periodic functions are similar to harmonic signals and have repeated values during each cycle, but a periodic function can have triangular, square, or any shape during the period. Therefore harmonic motion can be considered a special case of the periodic function. Fig. 1.21 illustrates a periodic function.

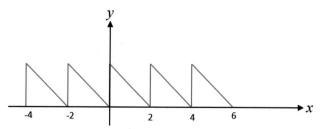

FIGURE 1.21 An example of a periodic function.

The nonperiodic motion is also known as transient motion when vibration has a periodic nature that can diminish with time.

Because of the cyclic nature of motion during vibration where the signal magnitudes are oscillating between positive and negative values, the average and mean of the vibration signals can have zero values, especially with sinusoidal signals and where there is no shift in the signal. In this case, the standard deviation of the signal becomes very similar to its RMS value. Because the RMS value is the sum of the squares of the values at different times, the negative values become positive and the RMS value is always positive. Therefore the RMS value is a very important and popular metric in vibration analysis and is used to represent the magnitude of the vibration signal.

Vibration motion can also be deterministic, nondeterministic, or random. Deterministic vibration signals are repeatable and can be predicted, meaning that by knowing the time, the value of the signal can be determined at this time or a future time. A sine wave is one form of a deterministic signal with known properties and behavior. An example would be riding on a swing. Fig. 1.20 shows an example of a deterministic signal. Nondeterministic vibration signals are random in nature and unpredictable, meaning that future values cannot be accurately predicted. Examples include earthquake signals and vibration resulting from driving on irregular roads (Fig. 1.22).

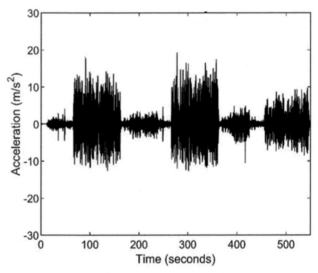

FIGURE 1.22   An example of the time-history of a random vibration signal at a vehicle floor moving across different terrains; the acceleration magnitude is represented by the vertical axis and time by the horizontal axis.

Due to their random nature, the description of random signals in the time domain is normally not a practical approach, and special treatments are required. Therefore nondeterministic signals can be described in terms of certain averages using probabilistic and stochastic approaches.

Vibration signals can be classified as stationary and nonstationary signals. A stationary signal's statistics do not change with time, but a nonstationary signal's statistics can change with time. In general, many real-life applications can be represented by stationary signals and can be analyzed and presented in terms of their statistical metrics.

Due to the uncertainty in a random process, experimentation is normally repeated many times, and collections of the vibration data in these repeated experiments are used to calculate the statistical average of the random process. For example, in testing human response to vibration, human subjects are exposed to vibration signals, and their excitation and response data are stored. Each experiment can be considered as a sample function. Because of the complexity of the human body, the human response in each sample function can be different. Therefore statistical averages are normally conducted on these sample functions to find the mean value of the random process. In this case, the mean value of a random process at a given time is normally found by summing the values of the random variables from each sample function and then dividing by the number of sample functions. Another type of function is the autocorrelation function, which can be calculated by summing the values of the product of two values from a sample function [at time (t) and after a short period $(t + dt)$] across the function time and then dividing them by the total time. The maximum value of the autocorrelation function is equal to the mean square value of the signal. The statistical average of a random signal can also be estimated using one representative sample function, preferably one with a longer time (Meirovitch, 1975).

The concept of shock can accompany vibration in many applications where the sudden application of a force can give the system a high acceleration value in a certain direction. One example is when a transport vehicle moves at a high speed on irregular terrain. In this case, extreme force can be generated and create energy that can cause extreme motion and potential damage. Shocks are transient physical excitations that can generate oscillatory vibrational motions. A sudden impact to a bridge or any mechanical system can generate a transient vibration that shakes the system and then dies out with time. The speed of returning to the original equilibrium stage depends on the amount of damping in the system. Normally, a stiffer system vibrates for a longer time than a less stiff, or compliant, system. In instances when shocks generate high acceleration that exceeds gravitational acceleration and forces that exceed the weight of the system, the impact of the shock will force a person sitting or lying in a vehicle to lose contact with the supporting surfaces.

1.7.1   Random vibration analysis

Due to the nondeterministic nature of motion, random vibrations use statistical or average characteristics of the motion of a randomly excited system. Several functions are traditionally used in vibration analysis, including autocorrelation, power spectral density (PSD), cross-correlation, transfer functions, and coherence functions. The transfer functions can be effectively used to represent linear systems; however, their representation and prediction of the system behavior and response become weaker when the system is nonlinear. A nonlinear system is any system that does not have linear characteristics. For example, a system with a sine or cosine response is a nonlinear system. The coherence function is a good measure of the degree of linearity of the system and a measure of the strong correlation between the input and output motions of the system. The coherence function has a value between 0 and 1, with 1 indicating strong linearity of the system and a strong correlation between the input and output. Values lower than 1 indicate weaker correlations and nonlinear behavior of the system. The PSD function provides information similar to that of autocorrelation but in the frequency domain instead of the time domain. PSD describes the distribution of the power over the frequency components of the signal and can be either zero or positive. This becomes an important tool when investigating the power distribution in whole-body vibration (WBV) signals and finding the critical frequencies with the highest power. Cross-correlation is similar to autocorrelation, but in this case, the product is between two signals, $x(t)$ and $f(t)$, in the time domain instead of multiplying the signal by itself for autocorrelation. The cross-spectral density between two signals represents the transformation of the cross-correlation to the frequency domain. The cross-spectral density is a very useful metric in random vibration as it is a complex function that has magnitude and phase information. The phase information is another important metric that quantifies the delay between the response of the two signals, the input and output, and how the delay is affected by damping.

1.7.2   Equivalent system

Because of the complexity of motion during vibration, original systems are usually simplified to an equivalent system. For example, to study the vibration of a building, each floor of the building is simplified to an equivalent system that comprises basic components such as beams. The human body with its complexity is sometimes simplified to a single vibrating mass. While many advanced commercial software programs are available to analyze the motion of very complicated systems, the idea behind equivalent systems is to gain a general understanding of the motion. In addition, analysis with an equivalent system may be sufficient for certain applications. The point here is that simplification is sometimes a powerful and low-cost approach. One of the simplest fundamental systems in vibration is the single-mass, single-degree-of-freedom system (Fig. 1.23).

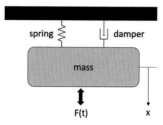

FIGURE 1.23   Single mass–spring–damper system.

The system comprises a mass, spring, and damper. The mass, in kg, represents the mass of the system, is normally rigid, and does not change in shape or magnitude during motion or under forces. The spring represents the flexibility of the system. In a stiff system, the application of a large force can induce small deformation in the spring, and with less stiffness or with compliant springs, the spring can show large deformation with smaller forces. The damper represents the damping in the system, such as the soft materials in the cushion of a seat or the suspension system in cars. When forces are applied to the system, they cause the system to vibrate; the mass will oscillate while the spring and damper work to resist, absorb, and dissipate the energy of vibration. This system is called a single-degree-of-freedom system because it has one mass moving in one direction (up and down). If the system has more than one mass or it moves in more than one direction, it is called a multidegree-of-freedom system. The degrees of freedom of the system depend on the number of independent variables that can be used to describe the motion of the system.

While a single mass–spring–damper system seems very simplistic, many researchers have simulated human response to vibration as a single mass–spring–damper system. This is especially true when the human body moves like one mass sitting in a seat and exposed to up-and-down vibration. For example, Wei and Griffin (1998) used a single mass–spring system and a two-mass, two-spring system to model seated humans under vertical vibration. The models were able to capture most of the human response in an acceptable manner.

The derivation of the equation of motion of a single mass–spring–damper system can be found in many vibration books (Newland, 1984), and conceptually it is an extension of Newton's second law with the addition of the forces that result from the resistance to the displacement (kx) and the damping ($c\dot{x}$), which is proportional to the velocity of the system ($\dot{x}$).

$$m\ddot{x} + c\dot{x} + kx = \mathbf{F(t)} \tag{1.43}$$

This equation is called the equation of motion, where m is the mass of the system, $\ddot{x}$ is the acceleration of the mass, c is the damping in the system,

$\dot{x}$ is the velocity of the system, k is the stiffness of the system, x is the displacement of the system, and F is the applied external force on the system. This equation can also be looked at as a balance between the applied forces and the system forces. In this case, $m\ddot{x}$ represents the inertial forces. The inertial force is everything that is related to the resistance of the mass and the shape of the system to the motion. The second term, $c\dot{x}$, is the damping force that works to dissipate the energy from the system, and the third term, kx, is the stiffness force or the resistance to deformation in the spring. While Eq. (1.43) is derived for a single mass, systems with many masses can be written in a similar manner, as follows:

$$\mathbf{M\ddot{x} + C\dot{x} + Kx = F(t)} \tag{1.44}$$

where $\mathbf{M}$, $\mathbf{C}$, and $\mathbf{K}$ become matrices instead of single numbers.

While the equation of motion can be solved analytically or numerically to determine the displacement, velocity, and acceleration of the system under the effect of external forces, many useful features can be derived from Eq. (1.43). One of these features is the natural frequency of the system (ω_n). The natural frequency, normally measured in Hz, or cycles per second, is a system natural property and will not change under forces. If the system is exposed to a frequency that matches its natural frequency, the system will experience large motion and can be damaged if is exposed to that frequency for a long time (resonance). At resonance, the energy entering the system will be trapped inside the system and will have a hard time leaving the system. The natural frequency of the system represents the ratio between its stiffness and its mass. The mathematical expression for ω_n of a single-mass, single-degree-of-freedom system is

$$\omega_n = \sqrt{\frac{k}{m}} \tag{1.45}$$

where k is the stiffness and m is the mass. In this case, heavy and compliant (small stiffness) systems have a low natural frequency, while stiff and light systems have a high natural frequency.

Systems with many masses and degrees of freedom have many natural frequencies. For example, systems with two masses moving in one direction are considered two-degree-of-freedom systems, as it is required to attach a coordinate system to each mass to describe their motions. Continuous systems like a beam or plate can have an infinite number of small masses connected via springs and dampers and therefore have an infinite number of degrees of freedom and natural frequencies. Each one of these frequencies, if excited, can generate significant motion. At higher natural frequencies, the systems demonstrate motions with small amplitude or displacement as compared to low frequencies that normally comprise larger displacement and slower time. Therefore in most engineering applications, especially with human whole-body motion, the interest is mostly in low frequencies generated from machines in work environments such as during transport, construction, and farming, which are normally less than 100 Hz. The human balance system operates with frequencies below 20 Hz. Higher ranges of frequency can be considered when dealing with tools working in these higher ranges. The natural frequencies are normally called the modes or the eigenvalues of the system.

At each natural frequency, the energy trapped in the system will force the system to take a certain shape, called the mode shape or the eigenvector. The mode shape is an expression that defines the ratio between the displacement across points on the object and a reference point when the system is excited at its natural frequency, so it is a dimensionless quantity and can be scaled up and down. The system can have as many mode shapes as the number of its natural frequencies. Each mode shape demonstrates the shape of the object under its corresponding natural frequency. Fig. 1.24 shows the first three mode shapes of a beam that is fixed at both ends, which can be used as a model of a bridge girder.

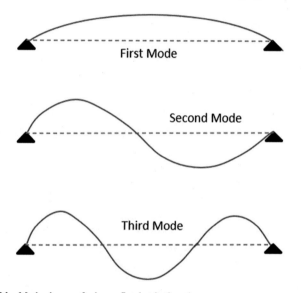

FIGURE 1.24 Mode shapes of a beam fixed at both ends.

Normally, the first mode will have most of the energy and the subsequent modes will have less energy and relatively less displacement. In addition, the number of nodes (stationary points) increases with higher modes. For example, the first mode in Fig. 1.24 has two nodes at its supports, while the third mode has four nodes, two at the supports and two at the intersection with the dashed line. The mode shapes shown in Fig. 1.24 are all bending modes because the beam is modeled in a plane; for three-dimensional structures such as a bridge deck, mode shapes include twisting, extension, and compression as well. The human body and its segments, as a biomechanical system, have many natural frequencies and can take different mode shapes.

1.7.3 Human modeling in vibration

In addition to the simplified single-degree-of-freedom human models, there have been many attempts to develop computer models of humans during vibration with different complexities. Approaches using the current state of the art in commercial finite element modeling and multibody dynamics software like Abaqus (Ataei & Mamaghani, 2018) are traditionally used. In addition, models that use several links to simulate the human body have been also developed. Many models have been developed for seated and standing humans (Buck & Wolfel, 1998; Fritz, 2005), but fewer models have been developed for supine humans (Qiao & Rahmatalla, 2019; Wang & Rahmatalla, 2013b).

1.7.4 Modeling of human head−neck

The human head−neck is a very complicated part of the human body. Simplified models, such as the one shown in Fig. 1.25, can provide good

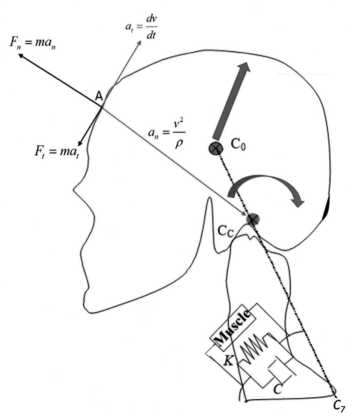

FIGURE 1.25 Planar model of a human head−neck showing motions and forces. *Adapted from Wang, Y., & Rahmatalla, S. (2013a). Human head−neck models in whole-body vibration: Effect of posture.* Journal of Biomechanics, 46, 702−710.

predictions of head—neck motion when subjected to external forces or motions under some circumstances. More detailed models, however, can provide more insight into what will happen between the head and neck when the relative motion between them is allowed to take place (Wang & Rahmatalla, 2013a).

In this planar model (Fig. 1.25), the head, with a center C_0, is modeled as a rigid body with a specified shape. The neck is also modeled as a rigid body and connects the lower part of the head at (C_c) and to the upper part of the torso at the seventh cervical vertebrae (C_7). The tissues and muscles surrounding the head—neck region are normally modeled as springs and dampers. The springs represent the stiffness, or resistance to deformation, and the damper represents the damping, or energy absorption, of this area. A muscle model, such as the one presented by Rahmatalla and Liu (2012), can include an active component that represents the forces the muscles can generate to resist motions.

When the head is exposed to motion, such as during transport, it will be exposed to two types of motion: a rotational component that rotates the head about its center of motion about C_c and a translational motion that works to pull/push the head from the location where it is connected to the neck. These types of motions, according to Newton's second law, will generate two types of forces: one force, F_n, will pull/push the head out of its connection joint (C_c), and the other one, F_t, will apply a moment and a shear force that tries to bend and shear the head at its connection joint (C_c). If the model is extended to the three-dimensional space so the head becomes like a sphere, a twisting motion can also take place and will generate a twisting torque on the joint (C_c). Of course, tissue and muscles will take most of the forces that can result from motion and absorb and dissipate them. Still, there are always limits on how much force the tissue and muscles can handle without causing damage to this region.

The following example presents a computer model that was developed for supine human transport in a WBV application (Qiao & Rahmatalla, 2019; Wang & Rahmatalla, 2013b).

1.7.5 Supine-human model

Until now, only a handful of supine-human models in response to vertical WBV in the gravity direction have been found in the literature (Peng et al., 2009; Vogt et al., 1978). Vogt et al. (1978) proposed a multimass system model that reproduced the experimental data of nine human subjects under sinusoidal vibration with a centrifugal constant acceleration magnitude. Their proposed model comprised four segments—the head, represented by a pure mass; the chest; the abdomen; and the legs—with each represented by a multidegree-of-freedom lumped parameters model. The motion of the masses of the different segments representing the body is not directly connected in this model, meaning that the mass representing the head will move up and down independent of the mass representing the torso or other masses. Therefore no motion coupling between the masses exists in this model. The study showed that the thorax region of the human body reacted differently than the rest of the body segments. This was explained by the thoracic anatomical configuration, where most of the body's organs reside. Another supine-human model was proposed by Peng et al. (2009) to predict the dynamic response of the human body when lying in a railway sleeper carriage under random track vibration. The model comprised a 14-degree-of-freedom human-berth-coupled dynamic system. It simulated the human body with three segments representing the head, buttock, and leg, as well as the underlying carriage system. One limitation of the latter models is their inability to capture the rotational motion of the human-body segments, as all segments are vibrating in one translational direction (up and down). Also, these models lack rotational joints between the body segments and therefore could underestimate the dynamic interactions between the adjacent segments.

More recently, a coupled-segment human model was introduced (Wang & Rahmatalla, 2013b) to simulate a supine human and the underlying transport system in response to vertical WBV. For this model, the human body was simulated by three segments: the head−neck, torso, and pelvis−legs. The segments of this model were connected using joints that were allowed to have translational and rotational motions. The transport system upon which the humans were lying was modeled using spring and damper elements (Fig. 1.26).

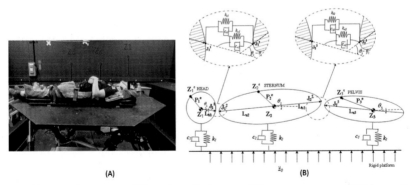

FIGURE 1.26 Computer modeling of a supine human and the transport system: (A) lab experimentation on real humans, and (B) a three-mass model of a supine human with the transport system.

When tested and validated with human data, the model showed reasonable results. Fig. 1.27 shows the predicted response of the model under WBV with random signals in the vertical gravity direction (Z-direction in the figure).

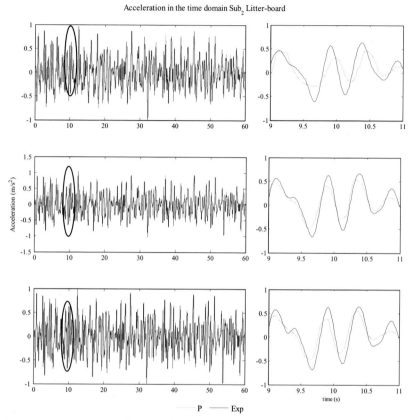

FIGURE 1.27 Experimental and predicted acceleration time history at the head, torso, and pelvis under the board-litter-case condition for a specific subject.

The figure shows that the model gave a reasonable prediction when compared with the data from human subjects. It should be mentioned that the data from seven supine human subjects exposed to random vertical vibration were used in identifying the parameters of the model. The data from a new subject were used to validate the model efficacy. In the latter process, anthropometrical and response data from the new subject were used in the validation process of the model.

1.7.6 Human coordinate system in whole-body vibration

The International Standardization Organization (ISO 2631-1, 1997) has adopted a body basicentric coordinate system (ISO 2631-1) that can be applied to humans taking different postures, including standing, seated, and supine positions. Fig. 1.28A depicts the orientation of supine humans.

Because the human-body segments can move and rotate in different directions during vibration, local coordinate systems may be attached at the head, torso, pelvis, and legs (Fig. 1.28B). These local systems can be rigidly attached to each body segment and are used to calculate the relative motion between the adjacent segments. The significance of the local systems can become essential when investigating the efficacy of immobilization systems such as a neck-collar or vibration mitigation system that target a certain location or different locations on the body.

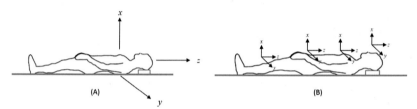

FIGURE 1.28   Body basicentric coordinate system (A) as described by ISO 2631-1; (B) local basicentric coordinate systems applied at the head, chest, pelvis, and legs.

1.8   Chapter summary

This chapter covers basic concepts in linear algebra, complex numbers, forces and motions of particles, rigid bodies (with examples on human models), basic statistics (including the meaning of the p-value and confidence intervals), and time and frequency analysis. It concludes with an introduction to vibration that includes vibration fundamentals, random vibration, and equivalent systems, including examples of supine humans. This chapter forms the basis for understanding the concepts that are utilized in the rest of the book. One point that should be taken from this chapter is that complicated motions and systems can be simplified, and simplification can be a very powerful tool. As demonstrated in this chapter, the random vertical vibration of a seated human can be successfully represented by a single-mass, single-degree-of-freedom system. Of course, systems with more masses and degrees of freedom can give better predictions and more in-depth analysis.

References

Ataei, H., & Mamaghani, M. (2018). *Finite element analysis: Applications and solved problems using Abaqus®*. United States (publisher not identified).

Bedford, A., & Fowler, W. (2008). *Engineering mechanics: Dynamics* (5th ed.). Pearson.

Buck, B., & Wolfel, H. A. (1998). Dynamic three-dimensional finite element model of a sitting man with a detailed representation of the lumbar spine and muscles. *Computer Methods in Biomechanics and Biomedical Engineering, 2,* 379−386.

Eide, A. R., Jenison, R. D., Mickelson, S. K., & Northup, L. L. (2018). *Engineering fundamentals and problem solving* (7th ed.). New York: McGraw Hill Education.

Fritz, M. (2005). Dynamic properties of the biomechanical model of the human body-influence of posture and direction of vibration stress. *Journal of Low Frequency Noise, Vibration and Active Control, 24*(4), 233−249.

International Organization for Standardization 2631-1. (1997). *ISO 2631−1. Mechanical vibration and shock—Evaluation of human exposure to whole-body vibration—Part 1: General requirements* (ISO Standard No. 2631−1:1997). https://www.iso.org/standard/7612.html.

Kreyszig, E. (1993). *Advanced engineering mathematics* (7th ed.). John Wiley and Sons, Inc.

Liu, Y. (2011). *Dynamics and control of mechanical and biomechanical systems* (unpublished doctoral dissertation). The University of Iowa.

Meirovitch, L. (1975). *Elements of vibration analysis*. McGraw-Hill, Inc.

Newland, D. E. (1984). *An introduction to random vibrations and spectral analysis*. Longman Inc.

Peng, B., Yang, Y., & Luo, Y. (2009). Modeling and simulation on the vibration comfort of railway sleeper carriages. In *International Conference on Transportation Engineering 2009* (pp. 3766−3771).

Qiao, G., & Rahmatalla, S. (2019). Identification of damping and stiffness parameters of cervical and lumbar spines of supine humans under vertical whole-body vibration. *Journal of Low Frequency Noise, Vibration & Active Control, 39*(1), 59−71. Available from https://doi.org/10.1177/1461348419837031.

Rahmatalla, S., & Liu, Y. (2012). An active head−neck model in whole-body vibration: Vibration magnitude and softening. *Journal of Biomechanics, 45*(6), 925−930.

Vogt, L. H., Mertens, H., & Krause, H. E. (1978). Model of the supine human body and its reactions to external forces. *Aviation, Space, and Environmental Medicine, 49*(2), 270−278.

Wang, Y., & Rahmatalla, S. (2013a). Human head−neck models in whole-body vibration: Effect of posture. *Journal of Biomechanics, 46*, 702−710.

Wang, Y., & Rahmatalla, S. (2013b). Three-dimensional modeling of supine human and transport system under whole-body vibration. *Journal of Biomechanical Engineering, 135*(6), 061010−061013. Available from https://doi.org/10.1115/1.4024164.

Wei, L., & Griffin, M. J. (1998). Mathematical models for the apparent mass of the seated human body exposed to vertical vibration. *Journal of Sound and Vibration, 212*(5), 855−874.

Chapter 2

Measurement of human response to vibration

2.1 Introduction

When evaluating the efficacy of transport and immobilization systems, it is vital to accurately measure the motion of the transport system and the motion of the human body relative to the surfaces of the transport systems during vibration. Accelerometers have long been the gold standard in whole-body vibration (WBV) studies, but recent advancements in sensors and sensing technologies have made it easier to measure human response in complicated environments such as WBV where the magnitude and direction can be changing randomly. This chapter focuses on measurements of the motion of the human body and the surrounding equipment. Other types of measurements, including forces and muscle activation, will not be discussed here; references will be cited if the topic is mentioned in the text. Although this book focuses on the response of humans in supine positions in WBV environments, examples of measurements of objects and standing and seated humans will be included to demonstrate some basic concepts.

2.1.1 Historical background

Major vibration studies on humans were launched during the early 1960s, triggered in part by interest on the part of NASA and the US Air Force in the effects of vibration on the human body during space transport (Coermann, 1961; Magid et al., 1960). More work followed, and vibration testing on humans using accelerometers has continued for the ensuing six decades.

To date, most vibration measurements have focused on the spinal area of the human body due to the perception that vibration exposure over a long period of time can cause chronic back pain for people who are seated while operating heavy construction or agricultural equipment. It is known that vibration enters the human body through the surfaces where it contacts the surrounding equipment, so measurements have mostly focused on that interface as the major source of the vibration energy entering the human body. In WBV, domestic and international guidelines/standards and European

Prehospital Transport and Whole-Body Vibration. DOI: https://doi.org/10.1016/B978-0-323-90103-1.00001-2

Commission laws dictate exposure limits based on the measurement of vibration at the interface between the seat and the operator's buttocks using seat-pad accelerometry (American National Standards Institute, 2002). For a seated human, most energy will come from the seat cushion, and therefore measurement of the input vibration to the human body is always done at the interface of the person's buttocks and the cushion. Similar concepts have been applied to supine and standing humans. During the early period of human testing, due to technological limitations and other practical challenges, most measurements were based on the energy entering the human body at the interface in a single direction, mostly in the vertical gravity direction, and not on what was happening to the human body. This was also reflected in international standards, such as the British Standards Institution (1974) and ISO 2631-1 (International Organization for Standardization, 1997), which were written during the mid-1970s. However, there has also been significant work during the last five decades to measure the biodynamics of humans, mostly in the spinal area, due to vibration, and many techniques have been suggested to measure the realistic motion of the spine.

It should be mentioned that ISO 2631-1 was introduced to provide guidelines and define methods for quantifying and evaluating WBV with regard to its effects on human health and comfort. While measurements of recumbent persons were introduced in ISO 2631-1, the analysis in the standard was in general based on the effect of vibration on the lumbar area of seated humans. That was historically done to answer questions regarding the potential of chronic back pain resulting from prolonged exposure to vibration such as that experienced by operators of heavy construction and agricultural equipment. In general, the standards are considered good documents in terms of defining standardized ways to evaluate humans under vibrational environments, which facilitates the comparison of data collected at different labs. The standards also define the location of the input and output measurements, the coordinate systems of the human body, details about the measurement locations, the general requirements for signal conditioning, and the duration of measurement. One limitation of ISO 2631-1 is that the measurements for the different evaluations are conducted at the interface between the human and the contacting surfaces and not on the human body. This type of evaluation may be appropriate for comfort and health evaluation, but it has limitations when considering measurements of human posture, the relative motion between the body segments, and the motion of the surrounding equipment. It is hoped that the material presented in this chapter will assist readers in conducting measurements and attaching sensors on the human body and the supporting surfaces.

2.2 Traditional measurement techniques in WBV

This section introduces the different types of sensors used to measure human motion in WBV environments. The International Organization for Standardization

(ISO) has published guidelines for the measurement and evaluation of human response under WBV (American National Standards Institute, 2002); interested readers may refer to the standard for more details.

2.2.1   Accelerometers

Accelerometers (Fig. 2.1) are electromechanical devices used to measure acceleration magnitude with time. These measurements may be *static*, like the constant magnitude of gravity, or *dynamic*, like movement or vibrations. Accelerometers can be manufactured with different shapes and sizes that can be as small as several millimeters. In physics, *acceleration* is defined as the rate at which velocity changes with time. Any increase or decrease in speed and any change in direction results in acceleration. According to Newton's second law, a dynamic force will be generated when a mass is exposed to a change in velocity, and the direction of the resulting force will be in the opposite direction to that of the mass acceleration. When a car accelerates at a constant rate, the people sitting in the car will feel a force pushing their bodies toward the seatback. When a car brakes suddenly and its speed reduces rapidly, a negative acceleration, or deceleration, will occur; here a force will be generated that makes the people sitting in the car lean forward. This sudden motion can be severe and can lead to injury under extreme braking conditions if a seatbelt is not used. A similar feeling can be observed when a car enters a curved road; in this case, the occupants of the car will be exposed to forces that push them to the side, in a direction pointing outward from the direction of the center of the curve of the road. These forces are called centrifugal forces. During vibration, however, the motion of the body becomes more complicated because the acceleration resulting from vibration is cyclic; it is not constant and can result in a mixture of acceleration and deceleration. As a result, the human body will be exposed to pushing and pulling forces that can cause discomfort and potential injuries under severe vibration conditions.

FIGURE 2.1   An example of a commercial accelerometer (Dytran Instruments Inc, n.d.).

2.2.2   AC and DC accelerometers

Accelerometers can be classified into two classes: AC-response and DC-response. AC-response accelerometers measure only dynamic acceleration, while DC-response accelerometers measure both static and dynamic acceleration. DC-response accelerometers are most suitable for measuring human motion under WBV because the amount of static acceleration tells us the angle of the human body with respect to Earth, and the amount of dynamic acceleration allows us to analyze how the body is moving. Also, DC accelerometers can give more accurate results than AC accelerometers under low-frequency motion, which can occur in most WBV applications, because its sensing unit can more accurately follow the slow-moving input than the sensing unit inside the AC accelerometers. An example of low-frequency motion is sitting in a swing that makes one cycle every 2 seconds (0.5 Hz). Most DC accelerometers use piezoresistive elements and produce resistance changes in the strain gages or capacitive types (Micro-Electro-Mechanical Systems technology) that are part of the accelerometer's sensing system. Most common AC accelerometers, on the other hand, use piezoelectric elements for their sensing mechanism. Under motion, a seismic mass inside the AC and DC accelerometers causes the piezoelectric or the piezoresistive element to produce an electrical output that is proportional to acceleration.

2.2.3   Limitations of accelerometers in WBV

Although accelerometers have long been the gold standard in WBV studies, they have difficulty capturing complicated human motions during WBV for a couple of reasons. The first, and like most sensors, and because attaching accelerometers to bones is so invasive, they are normally attached to the skin or to a rigid base that is then attached to the body. When accelerometers are attached directly to the skin, the mass of the accelerometers on the soft tissue can result in a bouncing local motion that can generate significant error during vibration measurements, especially at high frequencies. For example, if an accelerometer is attached to the chest of a person sitting in a high-frequency environment, it may register the motion of the skin and not the motion of the bones underneath it. Therefore it could appear that the person is experiencing severe motions while in fact, the accelerometer is bouncing around on the skin. While the real motion should be the motion of the bones or segments, decoupling and isolating the skin motion from the bone motions can be very complicated.

There were many attempts in the literature to attach the accelerometers directly to the bony areas of the human body. One popular example is the use of a bite bar to measure the motion of a person's head. A bite bar is a rigid bar with an extension piece that a person can bite down on and hold during the testing. Accelerometers are attached to the bite bar in a certain arrangement to capture the motion of the head in the required directions (Griffin, 1976). Fig. 2.2 presents a schematic representation of a bite bar.

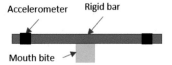

FIGURE 2.2   Schematic drawing of a bite bar; accelerometers are attached to a rigid bar, and a mouth bite allows a person to bite down firmly on the device, creating a rigid connection between the sensors and the skull.

While a bite bar can capture the real motion of the head, the weight of the bar and the way it moves during vibration can cause a lot of discomfort for the test subject. Some researchers have tried other techniques for attaching sensors to bony segments of the human body (Panjabi et al., 1986); in one example, sensors were attached to the spine vertebrae of volunteers using pins (Panjabi et al., 1986). However, these tests were conducted for scientific reasons for very special purposes and cannot be used in day-to-day experimentations.

The second issue with accelerometers is the difficulty of measuring motion on curved or irregular surfaces. A single-axis DC accelerometer, for example, is a device that can measure the acceleration in one direction only. If this accelerometer is placed on a flat surface (Fig. 2.3A) and there is no motion, the accelerometer will register a constant value of 9.812 m/s^2, which represents the magnitude of the gravitational acceleration. During up and down vertical movement, the accelerometer will register the magnitude of the acceleration due to movement in addition to the gravitational

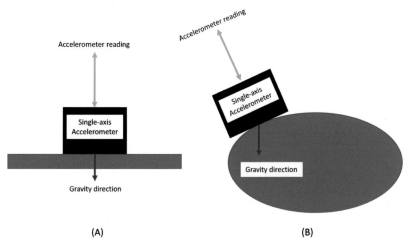

FIGURE 2.3   Attachment of a single-axis accelerometer on surfaces: (A) flat surface and (B) curved surface.

acceleration. In this case, the acceleration graph with time will have values reciprocating up and down around $-9.812 \, m/s^2$. Therefore the gravitational acceleration must be zeroed out by subtracting its value from the resulting acceleration, so the resulting acceleration from motion can be seen reciprocating up and down around zero and not around $-9.812 \, m/s^2$.

Dealing with the gravity components ($9.812 \, m/s^2$) becomes more involved when attaching the accelerometer to a nonflat or irregular surface as shown in Fig. 2.3B. In this case, it becomes more difficult to remove the gravity components from the real motion without knowing the magnitude of the inclination angle between the accelerometer and the curved surface with respect to the ground. Most measurement areas on the human body are irregular and nonflat, so when accelerometers are attached to these areas, that is, on the pelvis or legs, the process of attaching and aligning them with the direction of motion becomes difficult. It becomes even more complicated if the curved surface is exposed to rotational vibrations where the inclination angle may change with time—we will tackle this problem in more detail later in this chapter. Another issue with accelerometers is that, especially during multidirectional vibration, the body segments to which they are attached can move and rotate, and measurement directions can change over time. As a result, the accelerometers will measure motions in directions that differ from the direction that they are intended to measure. Under such conditions, methodologies that are capable of correcting the direction of measurements must be utilized.

Historically, accelerometers have been used to measure vibrational motion in a single direction, but they have recently been extended to measure motion in two and three directions. Translational motion (when all points of a body move in the same direction without rotation) can be measured with accelerometers, but the measurement of angular motion (when a body rotates) requires arranging a set of accelerometers on a rigid base at specific locations and orienting them to create one unit. In theory, measuring the three-dimensional translational and rotational motions (motion in space) of each segment of the human body would require six uniaxial accelerometers. However, due to the complicated relationship between the linear and angular parameters and the influence of gravity, between 9 and 12 accelerometers must be placed in a specific order and configuration to measure the complete translational (three directions) and rotational (three directions) motions of the resulting unit (Padgaonkar et al., 1975). Due to the size and weight of accelerometers, their use on humans in a WBV environment can be impractical. Therefore other means of measuring angular motion, such as marker-based motion capture systems and inertial sensors, are becoming more common.

2.3 Marker-based motion capture

A marker-based motion capture system, such as those produced by Vicon, Motion Analysis, and OptiTrack, is another way to measure human motion during WBV. Marker-based motion capture systems are widely used in biomechanical studies (Kurihara et al., 2002; Leardini et al., 2005), and their results have been proven to be accurate, repeatable, and consistent. Systems with a sampling rate that can exceed 200 frames per second and with a resolution that can reach 16 megapixels are available on the market.

In these systems, the markers are passive sensors, meaning that they are merely reflective surfaces and can be attached easily to any area of the body without the need for wires to connect them to a data acquisition system. Different sizes of markers can be used to track the motion of body segments, hands and fingers, and the face. The markers are normally attached to bony landmarks such as elbows, clavicles, and vertebral spinous processes. In gait analysis, for example, the markers are attached to the lower extremities and pelvic areas to measure walking style and walking abnormalities in patients. As the participant moves, the positional history of each marker is captured using an array of infrared cameras. Fig. 2.4 shows an example of one of these commercial cameras.

Each camera sends an infrared light, and when the light hits the marker surfaces, it reflects and is registered by the cameras. Based on these reflected lights, special algorithms are used to determine the precise locations of the markers relative to the cameras' coordinate systems. While the markers measure the movement of the body segment to which they are attached, the position of each marker in the lab space can also be identified with very high accuracy with respect to a fixed coordinate system in the lab.

Before collecting data for any project, the motion capture system normally goes through a calibration process composed of two segments. The first segment is called *static calibration*. In this process, a global coordinate system is established and fixed to the lab. The global coordinate system will

FIGURE 2.4 An example of a motion capture camera by OptiTrack (OptiTrack, n.d.).

FIGURE 2.5   L-Frame by OptiTrack (OptiTrack, n.d.).

not move during the experimentations, and the locations of the cameras with respect to this global system are well defined. During static calibration, a set of markers is arranged precisely on a device called an L-frame (Fig. 2.5), which comprises two perpendicular links fabricated with high precision. Traditionally, one link holds three markers while the other holds one. The L-frame is placed on the floor of the lab in a location where all markers can be seen by all cameras; it represents a global coordinate system that is fixed in the lab space. The motion capture system will register the location of the L-frame and the distances between the markers and will use optimization techniques to minimize the differences between the physical and predicted distances. With this process, the position of the cameras in the lab can be identified relative to this global system.

The second segment in the calibration process is called *dynamic calibration*. This process is used to ensure that the cameras can track the motion of the markers in the lab space, relative to the global system, with a high degree of accuracy. The wand is another calibrated system in a T-shaped unit (Fig. 2.6). Three markers are usually attached at precise distances to one link; the number of markers on the other link may vary.

A lab operator will hold the wand in the lab and move it as if painting a wall. It can take up to a couple of minutes for the operator to cover the

FIGURE 2.6   Wand by OptiTrack (OptiTrack, n.d.).

capture volume. Once this process is done, the motion capture system software runs an optimization problem. In this process, the system minimizes the errors between the real measurements and the predicted measurements. This process can be repeated until the resulting errors are very small. After finishing the static and dynamic calibration processes, the motion capture system is considered ready to go.

During the experiments, the motion capture system will capture the motion of all markers visible to the cameras and will run until the operator pushes the stop button. The data collection process is followed by the postprocessing process in which the data are checked. Due to the complexity of WBV motion, some markers may become occluded from the cameras and their paths may not be fully captured, that is, their trajectories may have gaps. Part of data processing is to make sure to fill these gaps. One way to do that is with information from surrounding markers. This is why, in many motion capture studies, people attach more markers than they need; these are called *redundant markers*. The postprocessing operation traditionally uses specialized software that comes with the motion capture system, but some laboratories postprocess the raw marker data using in-house software for the subsequent data analysis. From the resulting marker displacements, the motion of any point on the human body can be identified with respect to the global system with a high degree of accuracy and with an error that can be less than 0.1 mm. With this information, the relative translational and rotational motions between the body segments with respect to any location in the lab can be easily calculated.

To help users determine where to attach markers on the human body, there are standard protocols, such as the one offered by Plug-In Gait (Schwartz & Dixon, 2018). While suggested marker locations may differ between protocols, they are all designed to achieve optimal placement with as few markers as possible. Theoretically, only three markers are required to define the three-dimensional displacement, velocity, and acceleration of each body segment; however, redundant markers are sometimes used to provide the most accurate results and to fill in trajectories that may be missing from the camera scenes (Rahmatalla et al., 2008). One major advantage of a marker-based system is the possibility of using many markers to capture the position and orientation of the human body and the surrounding equipment with no additional costs and with a high degree of accuracy.

Marker-based motion capture systems do have some disadvantages, however. First, marker attachments will produce errors because of the soft tissue motion issue that was discussed in the accelerometer section. These errors can be mitigated, though, because the markers are much lighter than accelerometers. Marker clusters can be attached to body segments via a rigid base, or individual markers can be attached to the skin with double-sided tape. Skin motion can be reduced by wrapping the area of the body where the markers are to be attached with athletic tape, as shown in Fig. 2.7. The implementation of such techniques at the author's lab has produced very consistent results.

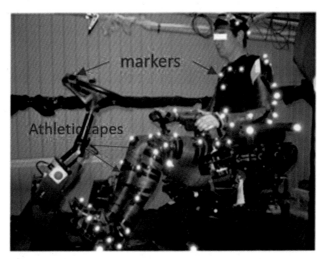

FIGURE 2.7 Markers attached to a seated human and the equipment surrounding him; the human is wearing a special motion capture suit (Velcro); athletic tape is added at the locations around the markers to reduce skin and relative movement.

Another disadvantage of marker-based systems in WBV applications is that they measure displacement rather than acceleration; mathematical operations, such as finite differences techniques, are then required to calculate the acceleration from displacement. Also, because of experimental and measurement noises, especially at high frequencies (above 16 Hz, or 16 cycles per second), the motion signal can become very small and nonsmooth, and the calculated accelerations, which require double mathematical differentiations of the displacement, can be erroneous. Experimental noise can result from the poor attachment of markers or the placement of markers on locations that encounter relatively high skin motions, while measurement noise can result from poor system calibration, markers with dirty or damaged surfaces, or the use of old, low-performance cameras. Nevertheless, the resulting errors can be mitigated by using state-of-the-art signal processing and filtering techniques. An additional issue in camera-based systems is that some markers can be occluded from camera view, which creates a gap in the measurement history and makes it difficult to measure the motion in certain locations at all times during the event. Occlusion events can be reduced by adding redundant markers as discussed earlier, but that should be done very carefully with WBV applications if a nonrigid relationship between the adjacent markers exists.

2.3.1 Velocity and acceleration from markers

In this section, a methodology is presented to determine the linear velocity, linear acceleration, angular velocity, and angular acceleration using markers. As indicated previously, the marker data can provide very accurate information about the position and displacement of selected points on the human body, especially when the connection between the markers and the body is rigid, meaning there is no relative motion or skin movement between the markers and the body. In WBV studies, however, acceleration is the traditional variable and is considered more desirable than velocity and displacement in most applications and standards. Therefore if markers are used for application in WBV, the displacement obtained from markers must be transformed to acceleration using mathematical expressions such as the finite difference method presented in Chapter 1, Fundamentals of Motion and Biomechanics.

Vibration signals can contain motions with very small displacements, especially at higher frequencies. While the motion at these high frequencies and low displacements can be easily captured by an accelerometer, difficulties will arise when using markers. A major difficulty is that the accuracy of markers can be reduced with smaller displacements. A second difficulty is the existence of experimental and measurement noise that can outweigh the magnitude of the displacement and create artificial and unrealistic acceleration. Therefore this section presents a case study to illustrate the validity and limitations of using markers' positional data to determine acceleration in WBV applications.

2.3.2 Methodology of using marker displacement to calculate acceleration

The goal of this section is to show the validity of using motion capture markers data to calculate the three-dimensional motion of a generic segment including linear velocity, linear acceleration, angular velocity, and angular acceleration. While this could be done for slow day-to-day motions such as walking and running, its application in WBV applications, where frequency components can reach high values, can become more complicated.

Fig. 2.8 is a schematic representation of a generic segment of the human body, the head as an example, as it moves between successive locations. Four markers are attached to the segment; three will be used to construct a local coordinate system rigidly attached to the segment, and one will be used as a redundant marker to fill in the gaps in adjacent markers if occlusion occurs. Additionally, one triaxial accelerometer is attached to the segment; it will be used for comparison with the acceleration calculated from the markers and for validation purposes.

Mathematically, the velocity components of any point on a rigid segment can be obtained as follows:

$$\mathbf{v}_s = \mathbf{v}_t + \boldsymbol{\omega} \times \mathbf{r}_{s/t} \tag{2.1}$$

where $\mathbf{v}_s$ is the velocity of a point (s) on the head segment, $\mathbf{v}_t$ is the velocity of another point (t) on the head segment, ω is the angular velocity of the

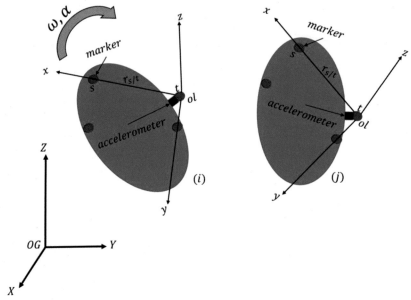

FIGURE 2.8   Schematic representation of a body segment during three-dimensional translational and rotational motions; OG represents the origin of a global coordinate system (lab coordinate system) and ol represents the origin of a local coordinate system rigidly attached to the segment.

head segment, and $\mathbf{r}_{s/t}$ is the relative distance between points t and s, which is assumed to have a constant magnitude. $\mathbf{r}_{s/t}$ can be calculated as the difference between the instant position of the marker at s and t. In Eq. (2.1), $\mathbf{v}_s$ and $\mathbf{v}_t$ can be calculated by differentiating the time history of the position of markers s and t between consecutive frames (i) and (j), or between any sequential frames. Once the linear velocities are determined, Eq. (2.1) can be used to solve for ω.

The acceleration components of any point on a rigid segment can be obtained as follows:

$$\mathbf{a}_s = \mathbf{a}_t + \alpha \times \mathbf{r}_{s/t} + \omega \times (\omega \times \mathbf{r}_{s/t}) \tag{2.2}$$

where $\mathbf{a}_s$ is the acceleration of point s, $\mathbf{a}_t$ is the acceleration of point t, and α is the angular acceleration of the segment. In Eq. (2.2), $\mathbf{a}_s$ and $\mathbf{a}_t$ can be determined using the central finite difference method between three or more successive frames of markers s and t. Eq. (2.2) can be then used to solve for α.

2.3.3   Case study of motion capture of seated subjects

In this case study, four markers were attached to each of three different body segments on seated human participants: the head, the torso, and the pelvis.

The participants were then exposed to WBV using a shaker table, which simulates the vibration of heavy construction and agricultural equipment during operations. The markers were attached to subjects' heads using a rigid halo, and the time histories of the marker locations were collected using a 12-camera Vicon motion capture system at a rate of 200 frames per second. Because of anticipated equipment and environmental noises, motion capture data are normally filtered to remove artificial motions from the data. For normal human motions, this is usually done with a low-pass filter (LPF) with a cut-off frequency of $8-10$ Hz. While the implementation of filters is critical to remove high-frequency artificial components, which look like small jumps over the main motion time history trajectory, filters can also remove important (real) motion from the motion trajectory. Therefore it is imperative to select appropriate filters and the right cut-off frequency for different applications.

For normal human motions like walking and running, the trajectories of the motion using markers can be anticipated and then filtered accordingly, but this is hard to do in WBV for many reasons. First, the motion in WBV is usually random and can vary from large-displacement, low-frequency motions to very small-displacement, high-frequency components; therefore the selection of an appropriate filter and cut-off frequency is challenging. Second, motions during WBV can contain many frequencies, and it is hard to tell which frequency is real and which is nonessential. To address these difficulties, filtering schemes based on spectral analysis techniques are used in WBV. Smoothing and differentiation techniques for motion data have been widely implemented in the literature (see, e.g., Rahmatalla et al., 2006). In this study, the popular technique of low-pass digital filtering with an 18 Hz cut-off frequency is implemented.

2.3.4 Results: validation of acceleration using accelerometers and markers

To validate the accuracy of the acceleration calculated from positional markers, the values are compared with the acceleration measured by the accelerometer. Methodologies for finding a local coordinate system for the accelerometer using markers will be detailed later in this chapter. The acceleration measured by the accelerometer and that calculated from the markers were compared in terms of frequency distribution and banded root mean square (RMS) acceleration around 4, 8, 16, 25, and 40 Hz (Fig. 2.9). As shown in the figure, filtering the marker data with LPF techniques worked well at frequencies up to 16 Hz, but the differences became larger at higher frequency bands.

2.3.5 Virtual markers

A major obstacle in using passive marker-based motion capture systems is occlusion, where some markers disappear from camera view during part or

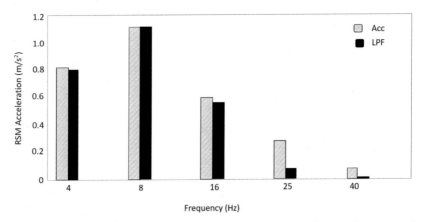

FIGURE 2.9 Comparison between root mean square acceleration of an accelerometer and acceleration based on markers filtered with a low-pass filter.

all of the testing. This becomes more critical when conducting studies where the measurements at certain locations are considered essential. One example is measuring the lower back motion of seated humans when that area of the body is covered by the seatback. One way to circumvent this problem is by using virtual markers. Virtual markers are traditionally used in human biomechanics, where a real marker is used at the beginning of the study to point to a certain location on the body. During the calibration process, a rigid-body relationship between this marker and the surrounding markers on the body can be established. After this process, the physical marker can be removed from the body, but its location and its relationship to the other markers are saved in the system. This means that the location of the marker, even after it is removed, can always be predicted from the surrounding markers. This predicted location of the marker after it is removed is called a virtual point. Normally, users can create more than one virtual point on the body to minimize the number of markers used during the experiment or to avoid locations where markers can be occluded from camera view.

2.3.6 Methodology of virtual markers

Fig. 2.10 shows a schematic drawing of the motion of a generic marker (P) in space. A global coordinate system with origin OG represents the fixed coordinates system in the lab setup. The local coordinate system with origin ol represents a local system that can be constructed from any three noncolinear markers that have a rigid-body relationship with P. While the global system is fixed and does not move, the local system is rigidly fixed to the body segment and moves and rotates with it in space. Point P represents a

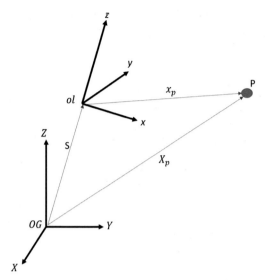

FIGURE 2.10 The position of a generic marker P, defined using a global coordinate system with origin OG and a local coordinate system with origin ol. *Adapted from Rahmatalla, S., Xia, T., Contratto, M., Kopp, G., Wilder, D., Frey-Law, L., & Ankrum, J. (2008). Three-dimensional motion capture protocol for seated operators in whole body vibration.* International Journal of Industrial Ergonomics, 38, 425–433.

generic marker that can move in space with respect to the global system but is considered to be rigidly attached to the local system.

When the marker P can be seen by the cameras, the global position of P can be calculated as follows:

$$X_p = S + Rx_p \qquad (2.3)$$

where X_p is the position of P with respect to the global coordinate system; and R is a transformation matrix between the local coordinate system and the global coordinate system and can be represented by the dot operation between the global and local coordinates.

$$R = \begin{pmatrix} x \cdot X & y \cdot X & z \cdot X \\ x \cdot Y & y \cdot Y & z \cdot Y \\ x \cdot Z & y \cdot Z & z \cdot Z \end{pmatrix} \qquad (2.4)$$

Similarly, the local position of P can be calculated as

$$x_p = R^T(X_p - S) \qquad (2.5)$$

where x_p is the position of P with respect to the local coordinate system and is rigidly fixed to the local coordinate system; S is the location of the origin of the local system with respect to the origin of the global system.

Because of the rigid-body assumption between the marker at P and the local coordinate system generated from the redundant markers, the distance x_p will be considered constant. This means that x_p is rigidly attached to the local coordinate system and will move and rotate with it during the motion.

In a case when the marker P is not seen by the cameras due to occlusion, Eq. (2.3) can be used to determine the global location of P during the experiment. While it is expected that the rigid-body assumption between the redundant markers and the physical markers cannot be maintained all the time because of the relative skin motion, the calculated x_p is considered an approximation of the physical position of the marker P; therefore the calculated distance x_p is considered to be approximately constant throughout the motion.

2.3.7   Case study of virtual markers

This case study (Rahmatalla et al., 2008) presents a situation where virtual points (markers) are used to track the motion of the upper back and the pelvis areas of a seated person during a WBV study. Fig. 2.11A shows a marker protocol that was designed to capture the motion of a seated subject during WBV. The marker protocol was designed to capture the motion of the upper body, including the head, arms, back, and pelvis. Under such conditions, the markers attached to the back and pelvis will disappear from camera view when the backrest is engaged (Fig. 2.11B). Once the backrest is engaged (Fig. 2.11B), the markers at T10, the sacrum level at the S1 vertebra, the left posterior superior iliac (LPSI) of the iliac bone, and the right posterior superior iliac (RPSI) of the iliac bone will disappear from camera view.

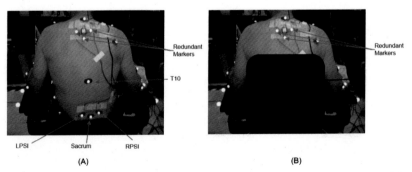

FIGURE 2.11   Positions of the physical markers (T10, S1, left posterior superior iliac, and right posterior superior iliac) and redundant markers on the subject's back when (A) the seatback is fully reclined and (B) the seatback is fully engaged. *Adapted from Rahmatalla, S., Xia, T., Contratto, M., Kopp, G., Wilder, D., Frey-Law, L., & Ankrum, J. (2008). Three-dimensional motion capture protocol for seated operators in whole body vibration.* International Journal of Industrial Ergonomics, 38, 425–433.

2.3.7.1 Data collection

The positions of the markers during the experiments were captured using a 12-camera Vicon motion capture system at a capture rate of 200 Hz. The resulting motion data were filtered using an LPF with a cut-off frequency at 16 Hz as described earlier in this chapter. While it is expected that the markers at T10, S1, LPSI, and RPSI will be occluded during normal working conditions when the seatback is in its engaged position, redundant markers were added to the subject's back and pelvis to create virtual markers (points) for T10, S1, LPSI, and RPSI. Fig. 2.11 shows redundant markers on a subject's upper back; they were used to create a virtual point at T10 when the seatback was engaged. Care should be taken when choosing the locations of the redundant markers to ensure that they have a rigid-body relationship with the virtual markers.

For the sake of conceptual analysis, the back/torso and pelvis areas of the seated person are schematically modeled by two ellipsoidal segments in Fig. 2.12. In order to create the virtual point at T10, three redundant markers are attached to the upper back. Additionally, three virtual points are required

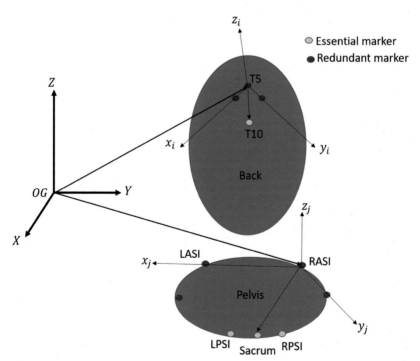

FIGURE 2.12 Schematic model showing essential (*green*) and redundant (*red*) markers. *LPSI*, Left posterior superior iliac; *RPSI*, right posterior superior iliac; *LASI*, left anterior superior iliac; *RASI*, right anterior superior iliac. *Adapted from Rahmatalla, S., Xia, T., Contratto, M., Kopp, G., Wilder, D., Frey-Law, L., & Ankrum, J. (2008). Three-dimensional motion capture protocol for seated operators in whole body vibration.* International Journal of Industrial Ergonomics, *38, 425−433.*

for the pelvis area at locations S1, LPSI, and RPSI. As shown in Fig. 2.11A, the addition of any number of redundant markers at the pelvis around S1 will be occluded by the seatback. In order to circumvent this issue and make use of the fact that the iliac bone is rigid, redundant markers are added to the left and right iliac crest and the left and right anterior superior iliac (LASI and RASI, respectively) as in Fig. 2.12 to ensure that they can be seen by the cameras while the seatback is engaged.

2.3.7.2 Results

Fig. 2.13 shows the displacement trajectories of the real and virtual point markers at S1 during 2000 frames. It is evident that the trends and magnitudes of the trajectories in the x- and y-directions are very similar and the error between the real and predicted trajectories is very small. While the trend in the z-direction is also good, there are some differences at the peaks. This error could be due to relative skin/bone motions that can change the rigid relationship between the redundant markers and the real markers. It could be reduced by enforcing the rigid-body assumption during the experiments, for example, by placing the redundant markers on wider bases.

2.4 Inertial sensors

Inertial sensors are based on inertia or mass properties and relevant measuring principles. They can have small mass and can measure translational and rotational motions in the three-dimensional physical space that we live in. Inertial sensors normally comprise an accelerometer that measures acceleration in the up-and-down, side-to-side, and fore-aft translational directions and a gyroscope that measures angular velocity in three rotational directions, namely pitch, roll, and yaw. The effectiveness of inertial sensors in correcting the direction of vibration can be enhanced with the addition of 3D magnetometers. With magnetometers, it becomes possible to measure the strength and direction of a magnetic field such as the north pole, adding another global reference to the measurement's orientations. With inertial sensors, it is possible to determine the position and orientation of the sensor in space. This information can be used to rotate and correct the direction of a sensor and make it align with the global direction or any other direction. This means that input and output measurements on the human body can be measured and aligned in similar directions, allowing accurate analysis in each preferred direction. This is very useful for multidirectional vibration, such as translational and rotational input vibrations encountered in the transport of a human on irregular terrain.

Inertial sensors could be especially beneficial to multiple-axis WBV studies where sensors' local coordinate system can be related to the global coordinate system, making it possible to investigate the relationship between the

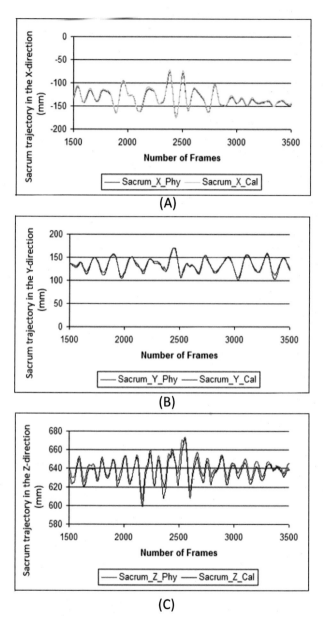

FIGURE 2.13 The trajectories of the sacrum marker for the second subject using physical (Phy) and virtual (calculated, Cal) markers in the (A) x-direction, (B) y-direction, and (C) z-direction for a whole trial of 7000 frames (35 s) and for only 2000 frames (10 s) in areas with significant activity. *Adapted from Rahmatalla, S., Xia, T., Contratto, M., Kopp, G., Wilder, D., Frey-Law, L., & Ankrum, J. (2008). Three-dimensional motion capture protocol for seated operators in whole body vibration.* International Journal of Industrial Ergonomics, *38, 425—433.*

motions of any segments of the body or/and the contact surfaces without worrying about the orientation of the sensors at these locations. While the capability to determine orientation is a major advantage of inertial sensors when compared to accelerometers, they can be sensitive to electromagnetic fields or metallic surfaces when they include magnetometers and may drift like traditional accelerometers over time. Magnetometers are vital to the ability of the sensor to provide absolute orientation values, but unfortunately, there is often equipment with metallic surfaces in WBV testing locations. These metallic surfaces interfere with the magnetic field and can generate errors in the calculated orientation of the sensors. This problem can be mitigated by modifying the surrounding surfaces with nonmetallic surfaces or by placing the sensors as far as possible from components that disturb the magnetometers' signals (Frick, 2015). It should be mentioned here that there are very efficient algorithms for predicting the orientation of inertial sensors when the magnetometers become corrupted (Frick, 2015).

One example of inertial sensor systems is Motion Tracking Sensor by Xsens (MTx) inertial trackers, in which the transformation matrix (orientation matrix) between the local coordinate system and a global system can be defined (Meusch & Rahmatalla, 2014). This provides an efficient way to transform the measurements from the local systems to the global system or between the local systems. Fig. 2.14 shows an example of these sensors and how they can be attached to the human body; the orientation of sensors attached to one segment of the body can be related to other segments to identify their relative orientations. Like other sensors, inertial sensors can be attached to the human body using double-sided tape or special attachments such as the halo shown in Fig. 2.14. Inertial sensors with either wired or wireless connections are widely available on the market. Like other motion capture systems, inertial sensors can be purchased with a suit that has special locations for each sensor on the human body. Most systems also come with software for data collection and postprocessing operations.

2.4.1 Transformation matrices from inertial sensors

In addition to measuring the acceleration and the angular velocity in three-directional space, most commercial inertial-sensor systems such as those by Xsens (2009) produce rotational matrices ($\mathbf{R}$) that can be used for the transformation of the resulting acceleration from the local coordinate system to the global coordinate lab-fixed system.

$$\mathbf{R} = \begin{bmatrix} q_1^2 + q_2^2 - q_3^2 - q_4^2 & 2(q_2 \times q_4 + q_1 \times q_4) & 2(q_2 \times q_4 + q_1 \times q_3) \\ 2(q_2 \times q_4 + q_1 \times q_4) & q_1^2 - q_2^2 + q_3^2 - q_4^2 & 2(q_3 \times q_4 - q_1 \times q_2) \\ 2(q_2 \times q_4 + q_1 \times q_3) & 2(q_3 \times q_4 - q_1 \times q_2) & q_1^2 - q_2^2 - q_3^2 + q_4^2 \end{bmatrix}$$

$$(2.6)$$

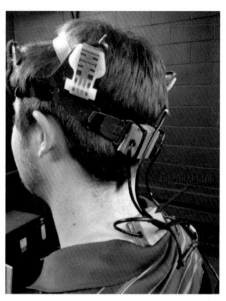

FIGURE 2.14 Inertial sensors attached to a halo firmly attached to the human head and the seventh cervical spine vertebra (C7) using double-sided medical tape.

where $q_1, q_2, q_3,$ and q_4 are the quaternion components of the motion. Users who are not familiar with quaternions can still use the matrix **R** as it will be presented in terms of values and not equations. The linear acceleration in the local sensor directions, local linear acceleration (a_x, a_y, a_z), can be transformed to the global system (a_X, a_Y, a_Z) as follows.

$$\begin{bmatrix} a_X \\ a_Y \\ a_Z \end{bmatrix} = \begin{bmatrix} R_{11} & R_{12} & R_{13} \\ R_{21} & R_{22} & R_{23} \\ R_{31} & R_{32} & R_{33} \end{bmatrix} \begin{bmatrix} a_x \\ a_y \\ a_z \end{bmatrix} \tag{2.7}$$

In addition to the transformation of the acceleration vectors between the local coordinate system of the sensor and the global directions of the lab, transformation matrices (**R**) between different sensors can be also established. This can be very beneficial when studying the relative motion between adjacent segments on the human body or between sensors attached to the human body and sensors attached to the surrounding equipment in the testing environment.

2.4.2 Case study: removing gravity components from accelerometer measurements

The gravity component (g), which represents the gravitational components of 9.812 m/s^2 in DC accelerometers, is part of the hardware. The gravity

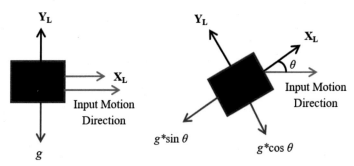

FIGURE 2.15 Orientation of an accelerometer where X_L and Y_L are the local axes of the accelerometer where (A) the g component is easily known and (B) the g component depends on inclination angle θ. *Adapted from DeShaw, J., & Rahmatalla, S. (2012). Comprehensive measurement in whole-body vibration.* Journal of Low Frequency Noise, Vibration and Active Control, *31(2), 63–74.*

component, which is a constant value, can be always mixed with the acceleration components that result from real motion. The gravity component can be easily removed from the accelerometer reading by simply subtracting it from the acceleration, especially if the sensor moves up and down in the gravity direction. Fig. 2.15A shows an example where the accelerometer is attached to a flat, horizontal plane. In this case, the g component is in the vertical gravity direction and has no effect on the acceleration in the x-direction, so the reading in the x-direction will not be affected by g. The g component can be subtracted from the vertical component of the acceleration if the motion is in the vertical direction.

The removal of the gravity component can become more involved when the sensor is attached to inclined surfaces and when a dynamic motion is involved. For planar motions, the g term will depend on the magnitude of the angle of inclination (Fig. 2.15B) and will have two components, one in the x-direction and one in the y-direction. When the inclination angle is zero (Fig. 2.15A), the x-gravity components become zero and the $y-g$ components become 9.812 m/s². The problem becomes more involved if the accelerometer has two planes of inclinations in space when attached to a spherical surface; for example, in this case, the acceleration from gravity will have three components in the x-, y- and z-directions.

The following algorithm can be used to remove the gravity component from the accelerometer reading when the accelerometer is part of the inertial sensor. The three accelerometers, three gyroscopes, and three magnetometers in the inertial sensors can be used to determine the absolute orientation values, therefore making it possible to extract the real acceleration signal and isolate the gravity components. The inertial sensors used in this work[1]

---

1. Xsens Technologies, Enschede, The Netherlands.

have a transformation matrix between the right-hand local coordinate system and a global system, as shown in Eq. (2.6).

The gravity components ($g = -9.812$ m/s^2 in the global Z-direction) can be calculated in each local direction by multiplying it by the transformation matrix **R**. The gravity contribution can then be removed from the inertial sensor's acceleration measurement in the local coordinate system, as follows:

$$\begin{bmatrix} a'_x \\ a'_y \\ a'_z \end{bmatrix} = \begin{bmatrix} a_x \\ a_y \\ a_z \end{bmatrix} + \mathbf{R} \begin{bmatrix} 0 \\ 0 \\ g \end{bmatrix} \quad (2.8)$$

where a_x, a_y, and a_z are the local raw-acceleration components and a'_x, a'_y, a'_z are the local accelerometer measurements where the gravity is removed. During vibration, this calculation can be performed at each time step of the motion.

Although the accelerometer measurements can be corrected by removing the gravity component, its measurement is in its local coordinate system. If it is desired to transform the measurement from the local coordinate system to the global coordinate system, then

$$\begin{bmatrix} a_X \\ a_Y \\ a_Z \end{bmatrix} = \mathbf{R}^{-1} \begin{bmatrix} a'_x \\ a'_y \\ a'_z \end{bmatrix} \quad (2.9)$$

where a_X, a_Y, and a_Z are the acceleration measurements in the global coordinate system with the gravity component removed.

2.5   Introduction to the concept of a hybrid system

As discussed earlier, most existing sensors have limitations in measuring the different components of motion during WBV. While accelerometers are the preferred measurement system for WBV, they have several limitations, such as their incapability to directly measure the orientation of the segment to which they are attached. In addition, more work is required to take the gravity components out of the resulting acceleration. Inertial sensors may be one of the best options, but they have issues dealing with drift and constructing accurate rotational transformation matrices when magnetometers are not used. Even with the addition of magnetometers, inertial sensors can be affected by surrounding metals and magnetic fields. Other sensors, such as motion capture passive markers, also have issues in terms of their incapability to directly measure acceleration, lower accuracy when measuring small displacement, and sensitivity to noise. Because all sensors have pros and cons in different applications, combining them in a *hybrid system* can achieve maximum benefit. For example, if accelerometers are combined with markers, the accelerometers can accurately measure acceleration while the markers can accurately measure the position and orientation of the accelerometers.

2.5.1 Hybrid marker–accelerometer system

FIGURE 2.16 A hybrid system with a triaxial DC accelerometer and four passive markers ($M1-M4$); $M0$ is a virtual marker calculated at the center of the accelerometer.

Fig. 2.16 shows an example of a hybrid system in which passive markers are attached to an accelerometer in a certain order to generate a local system at the center of the accelerometer. The accelerometer can be a single- or multiple-degree-of-freedom sensor. In this case, the accelerometer can accurately measure the local acceleration in a specified direction while the markers establish a local coordinate system that is rigidly attached to the accelerometer and moves/rotates with it. The sampling rate of the accelerometers is usually set to be more than two times the Nyquist frequency (the frequency below which the regenerated signals could be distorted). However, in practice, it is traditionally set to be more than 10 times the highest frequency, at a level of 500 frames per second for vibration testing with a maximum frequency of up to 50 Hz. The sampling rate of the markers can be 500 frames per second or more; however, most current commercial motion capture systems work at a rate of 200 frames per second. The local coordinate system from the markers can be used to determine the accelerometer's orientation relative to the global system or to any other system (e.g., the human head relative to the torso). Also, by using local coordinate systems, the acceleration information from the sensor can be projected/corrected in the intended directions (DeShaw & Rahmatalla, 2012).

As shown in Fig. 2.16, four passive markers are rigidly attached to the body of a triaxial DC-response accelerometer.[2] Only three markers are required to generate a local coordinate system; the fourth is a redundant marker. Using a 12-camera Vicon motion capture system,[3] the motions of the markers can be captured at a rate of up to 200 frames per second. The motion capture data are then synchronized with the accelerometer data at 200 Hz, meaning that the acceleration data will be resampled to be 200 frames per second instead of 500 frames per second. The next step is to assign a local coordinate system to the resulting hybrid system. Because the markers are placed orthogonally to (at right angles from) one another, the

---

2. Dytran 7523A1, Chatsworth, CA, United States.
3. Vicon, Los Angeles, CA, United States.

orientation vectors can be determined by simply subtracting one point in space (the center of one marker) from the center of another marker along each of the three directional axes. A virtual point can be calculated on the line connecting the markers and can be positioned at the center of the accelerometer by dividing the distance $M1-M2$ by 2. A local coordinate system can be constructed at the center of the accelerometer using a cluster of markers. For example, in Fig. 2.16, $M0$ can be selected as a virtual marker at the center of the accelerometer (Schwartz & Dixon, 2018).

In this case, the subtraction of markers $M4-M0$ can yield the X-axis vector, the subtraction of markers $M2-M0$ can yield the Y-axis vector, and the subtraction of markers $M3-M0$ can yield the Z-axis vector.

While these three vectors are supposed to be orthogonal and represent a local coordinate system of the accelerometer, there is a chance that, due to experimental errors, the constructed vectors are not perfectly perpendicular to each other. Therefore additional mathematical steps must be taken to ensure the orthogonality of the created local vectors.

In this process, the center of the coordinate system at the center of the accelerometer ($M0$) will be constructed from the intersection of the line between $M1$ and $M2$ and the perpendicular line from $M3$. A vector, $V1$, will start from $M0$ to $M2$ and point in the y-direction (Fig. 2.17). Another vector, $V2$, will start from $M0$ to $M3$ and point in the z-direction (Fig. 2.17), as follows:

$$V1 = \text{Coordinates of } M2 - \text{Coordinates of } M0 \qquad (2.10)$$

$$V2 = \text{Coordinates of } M3 - \text{Coordinates of } M0 \qquad (2.11)$$

$V1$ and $V2$ are usually normalized by dividing them by their lengths (the norms of $V1$ and $V2$, respectively).

$$V1' = \frac{V1}{\|V1\|} \qquad (2.12)$$

$$V2' = \frac{V2}{\|V2\|} \qquad (2.13)$$

where $\|V1\|$ and $\|V2\|$ are the norms of $V1$ and $V2$, respectively.

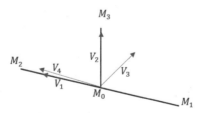

FIGURE 2.17   Schematic representation of the creation of an orthonormal coordinate system.

Using the right-hand rule, a cross product between $V1$ and $V2$ will result in a vector, $V3$, that is perpendicular to the plane containing $V1$ and $V2$ and that is pointing in the x-direction (Fig. 2.16), where:

$$V3 = V1 \times V2 \qquad (2.14)$$

This process will be followed by another cross product between $V2$ and $V3$ to produce a vector, $V4$, perpendicular to the plane containing $V2$ and $V3$.

$$V4 = V2 \times V3 \qquad (2.15)$$

This new vector, $V4$, will point in the y-direction (Fig. 2.16). If everything was perfect during the installation of the markers on the accelerometer, then $V4$ will be aligned in the direction of $V1$. Nevertheless, the mathematical correction process could still be necessary as it guarantees orthogonality. Because the vectors were normalized, the resulting coordinate system is called orthonormal basis because it comprises three orthogonal unit-vectors: $V2$, $V3$, and $V4$.

A similar technique can be used to create a global (stationary) coordinate system at the location of the L-frame that was used during the static calibration of the system. Once a local coordinate system is established for each accelerometer, these local systems can be used to measure the location and rotation of the accelerometer with respect to the global coordinate system and with respect to any other accelerometer in the lab space. Also, each accelerometer can now measure local acceleration at the point where it is attached and can provide the direction of this measurement in the local and global coordinate systems.

The transformation matrix between the right-hand local coordinate system and a global system can be defined using the following transformation matrix:

$$\mathbf{R} = \begin{bmatrix} \mathbf{x} \cdot \mathbf{X} & \mathbf{x} \cdot \mathbf{Y} & \mathbf{x} \cdot \mathbf{Z} \\ \mathbf{y} \cdot \mathbf{X} & \mathbf{y} \cdot \mathbf{Y} & \mathbf{y} \cdot \mathbf{Z} \\ \mathbf{z} \cdot \mathbf{X} & \mathbf{z} \cdot \mathbf{Y} & \mathbf{z} \cdot \mathbf{Z} \end{bmatrix} \qquad (2.16)$$

where $\mathbf{X}$, $\mathbf{Y}$, and $\mathbf{Z}$ are the unit vectors in the global system and $\mathbf{x}$, $\mathbf{y}$, and $\mathbf{z}$ are the unit vectors in the local system. Next, the contribution from gravity needs to be determined by multiplying the transformation matrix $\mathbf{R}$ with the magnitude of gravity using Eq. (2.8). This calculation can be performed at each time step of motion. While the accelerometer measurements are corrected without the gravity component present, its measurement is still in its local coordinate system. Finally, if desired, the local coordinate system can be transformed using the inverse of the transformation matrix $\mathbf{R}$ to transform to the global coordinate system as shown in Eq. (2.9).

In order to test the validity and accuracy of this hybrid system under different motion conditions, the system will be tested under different static and dynamic trials as shown in the following sections.

2.5.2   Static testing of hybrid systems

A straightforward way to test the hybrid system in a static environment is to attach it to an apparatus where the sensors can be positioned in different known orientations using different off-alignment scenarios. A possible testing block is shown in Fig. 2.18. The block is stationary to allow all three translational axes to be off alignment with respect to a global coordinate system. In this example, the motion capture data and accelerometer data were both synchronized at 200 frames per second.

In this static test, if the accelerometer is sitting flat on one of its sides, the acceleration measurement values would register 9.812 m/s^2 (in the gravity direction) and zero in the other two lateral directions. When the accelerometer is on an inclined surface as shown in Fig. 2.18, each axis of the triaxial accelerometer will show a measurement value other than zero, and the acceleration value at each axis is not centered on zero. This occurs because, under a static no-motion condition, the accelerometer measures only the gravity acceleration vector, which is not aligned with any of the accelerometer axes. Because the acceleration is local and the inclination is arbitrary, one cannot tell the orientation or the true value of the acceleration (except for gravity) from this data. Fig. 2.19A shows that the static signals measured around 7.5, −6.0, and −4.5 m/s^2 in the X, Y, and Z channels, respectively. Fig. 2.19B shows the static accelerometer measurement with gravity removed using the hybrid motion capture system, and the acceleration in all accelerometer axes is close to zero.

FIGURE 2.18 Testing block for acceleration tests where all three accelerometer axes were rotated off alignment from the global coordinate system; the testing block was rigidly attached to the motion-shaking platform.

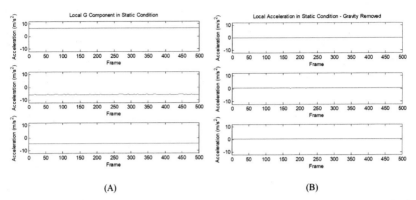

FIGURE 2.19 (A) Accelerometer signals for static condition on the testing block and (B) local accelerometer data with the gravity component removed for the static testing condition; note how each signal is close to zero.

2.5.3 Dynamic testing of a hybrid system

A six-degree-of-freedom man-rated motion platform (the Moog-FCS 628 electrical system) was used to test the hybrid method under dynamic conditions. The block shown in Fig. 2.18 was used again with the goal of keeping all local and global axes off alignment. The block was attached rigidly to the shaking platform. For each dynamic test, the shaking platform produced random vibration in the global x-axis for 10 seconds, random vibration in the y-axis for 10 seconds, random vibration in the z-axis for 10 seconds, and random three-dimensional (x-, y-, and z-axes together) vibration for 10 seconds. Data was collected at 200 frames per second for all accelerometer measurements and synchronized with the motion capture data, which was also collected at 200 frames per second.

Fig. 2.20A shows the raw local acceleration data of the dynamic vibration test, where the acceleration signals do not start from zero. Fig. 2.20B shows the local acceleration of the dynamic vibration test with the gravity component removed from the signal using the same methodology used in static testing so that the acceleration signals all now start from zero. In this case, the acceleration values started from zero in all directions. Similar corrections were also seen when the system was tested under vibration in a single direction (x or y or z).

Fig. 2.21 shows the global acceleration signal after being transformed to the global coordinate system. The figure illustrates how each signal is now oriented with the global reference: frames $0-2000$ show x-direction global vibration, $2000-4000$ show y-directional global vibration, $4000-6000$ show z-directional global vibration, and $6000-8000$ show 3D vibration.

2.5.4 Case study: simulated real-life application of hybrid system

After the static and dynamic proofs of concept, the hybrid system was tested in a simulated real-life application in which an operator conducted manual

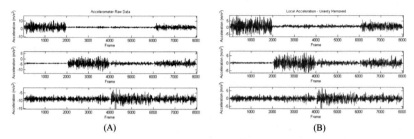

FIGURE 2.20   (A) Raw local accelerometer data in m/s² for x, y, z, and 3D vibration—note how each signal is not centered on zero and (B) local accelerometer data in m/s² for x, y, z, and 3D vibration with the gravity component removed—note how each signal is now centered on zero.

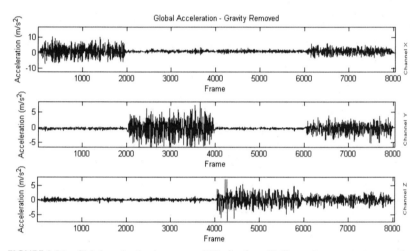

FIGURE 2.21   Global acceleration in x, y, z, and 3D vibration with the gravity component removed.

drilling using different drilling techniques. The measurement of motion in this application is considered very challenging because the operator is subjected to vibration from the tools while changing posture during the operations. In this case, the accelerometer will read the acceleration in the directions of its local coordinate system. It will be very difficult to transform these measurements in other directions, such as those required for human biomechanics, posture analyses, and human hand-arm vibration analysis, because those measurements must be aligned in certain directions, such as the direction of the bone of the forearm. With the hybrid system installed at the hand, forearm, upper arm, and shoulder, it would be possible to describe the posture, movement, and vibration of the whole arm and specify the acceleration levels at different locations (Fig. 2.22).

The hybrid system can be also applied to capture the whole-body posture of humans while they are conducting tasks where vibration is a dominant part of the environment. Fig. 2.23 shows an example of a person with a hand

FIGURE 2.22 Hand-arm vibration, posture, and movement application using a hybrid system.

FIGURE 2.23 Measurement of human motion, posture, and hand-arm vibration during a drilling task.

drill. In such situations, a marker-based system can be used to capture the motion and the posture of the person's whole body under different scenarios, including those when the person changes postures, while the hybrid system can be installed only at locations where the magnitudes and directions of the vibration need to be accurately measured. Grip force sensors can be added if needed, and the locations and directions of the forces can be also determined by adding three markers at the locations where the forces are measured.

2.5.5 Case study: measurement of a supine human under whole-body vibration

Measurement of the biodynamical response of supine humans due to WBV is essential for the evaluation of motion transferred to the human body and for testing the effectiveness of transport and immobilization systems. Understanding human response will provide crucial information to the industry to facilitate the development of efficient systems with vibration−mitigation properties. Most importantly, information about how humans respond will help medics find creative ways to support parts of the human body during prehospital transport, especially at locations where injuries may exist. An example is presented in this section to show how measurement is conducted on a supine human subjected to single-axis (e.g., up and down in the vertical gravity direction) and multiaxis (e.g., up-and-down and side-to-side directions simultaneously) WBV.

This section will first illustrate the equipment and the preparation of subjects for testing.

2.5.6   Motion platform (shaking table)

A man-rated motion platform (Fig. 2.24) is normally used to generate the vibrational motions for an experiment. It is designed so the shaking table cannot exceed the acceleration of gravity in any direction. It also has many features to ensure human safety during testing. In this example, a six-degree-of-freedom Moog-FCS628-1800 electrical motion system was used. This system can generate movements of greater than 0.39 m in the three translational axes and more than 23 degrees in the three rotational axes, with an accurate frequency response of up to 30 Hz. The accuracy of the signal reproduction was within 2% RMS. Additionally, the actual measured RMS accelerations between subsequent trials were also within 2% error.

The shaking table is an environmental simulation tool upon which rides collected from many machines or environments can be replayed. For example, a researcher can install motion sensors on the floor of an ambulance

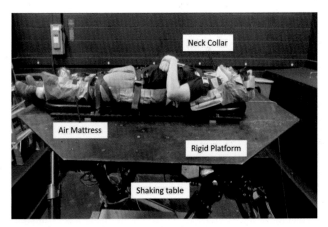

FIGURE 2.24   Shaking table at Iowa Technology Institute.

moving in a real-life situation. The researcher can then return to the lab and replicate reality. The data collected from the ambulance can be reproduced on the shaking table to simulate the ambulance environment. A human immobilized in a medical transport system can then be attached to the shaking table platform to simulate a person being transported by an ambulance (Rahmatalla et al., 2018, 2019).

In addition to the ride files that can be collected from real-life scenarios, the shaking table can generate and simulate a wide range of applications. For example, random input vibrations can be used with frequencies between 0.5 and 25 Hz of broadband Gaussian white noise in three directions to create hypothetical environments with different types of driving or flying (in the case of aerial transport) terrains. The broadband Gaussian white noise represents a signal that contains all frequencies with normal distribution. Such a range of frequencies is expected to capture dominant resonant frequencies existing in most ground vehicles. The length of the ride files can be arbitrary or can be dependent on the scope of the experiments. During lab testing, 60-second files will capture most of the activity in different scenarios, but longer files can be generated if required to simulate all ride events. Vibrations in any other form, as well as shock ride files, can be added in each of the three directions. Shocks can also be created for the study [e.g., see work by Mansfield (2017)] and can be random at unequal time intervals. The random 3D vibration files can be designed to offer approximately equal RMS accelerations (acceleration with equal magnitude) in all directions, allowing for the characterization of the whole system using one file. These lab-generated files are advantageous to the study because they include all aspects of a full band of vibration. While testing vibration due to foot, ground vehicle, or aerial transport may be good for specific applications, testing broadband vibrations covers a larger range of human body responses and transport scenarios.

2.5.7 Subject preparation

During prehospital transport, supine humans are normally placed on supporting surfaces that are quasirigid (e.g., spinal boards) or rigid with cushioned surfaces. These supporting surfaces can be attached to a stretcher or litter for easy handling while moving patients and loading them into transport vehicles. The stretcher or litter can then be firmly attached to the vehicle floor at specified locations. It is important to note that there are transport systems with different designs. For example, in Fig. 2.24, an air mattress is used instead of a spinal board to support the subject. Also, different techniques are used in civilian and military environments to immobilize patients; we will introduce some of them here.

In lab testing, the rigid platform of a shaking table represents the floor of the transport vehicle and moves in a similar manner. The transport system, which includes the litter/stretcher and the spinal board, is firmly attached to the shaking table platform. To begin the experiment, the human subject lies down on the spinal board, and a set of straps is used to secure the subject. The spinal board with the subject is then laid on the litter. A second set of straps is used to hold the litter to the person and the spine board, and a neck collar is normally used when transporting patients with suspected spinal cord injury (Fig. 2.24). In addition, the head is traditionally supported with a cushion underneath and two cushioned blocks on the sides to keep it from moving around. Finally, the litter is attached firmly to the platform of the shaking table.

Fig. 2.25 shows a person immobilized with the United States Army stretcher and immobilization system. The stretcher is fixed to the shaking table platform using clamps to ensure there is no relative motion between the stretcher and platform. In this immobilization system, a strap is placed at the bottom of the sternum, near the xiphoid process; another is placed across the pelvis; and a third is placed on the legs just above the kneecap. When securing the straps, it is important to use the appropriate amount of tension to reduce the motion of the patient while also reducing discomfort and allowing the patient to breathe normally. In practice, emergency responders use different subjective approaches, such as making sure two or three fingers can fit between the patient's body and the straps. In this example, force scales are used to illustrate another approach to maintaining prescribed tensioning across all straps. The force scales are attached to all free ends of the straps, and tension is applied until the scales show a reading of approximately 135 N. The 135 N measure was chosen based on preliminary tests to ensure subject comfort and security.

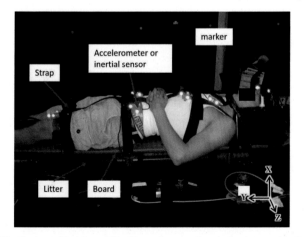

FIGURE 2.25   Supine subject immobilization and preparation during transport testing.

Fig. 2.25 also illustrates the protocol for a stretcher with a spinal board.[4] It is the same as the simple stretcher protocol but has an additional strapping system[5] to secure the subject to the spinal board. Straps in the spinal board system run through the spinal board and wrap around the subject at the knees, hip, chest, and shoulders. The same tension measures are used to join the spinal board straps; each strap is pulled through the medical backboard, over the subject, and then connected to the Velcro. Shoulder straps are secured by applying tension in the vertical direction and then connecting the strap to the Velcro. A pad included with the board system is used under the head to reduce discomfort on the bare plastic of the board. It is securely strapped to the board, and it is not preventing the head to move freely during the simulations. Additionally, the head is supported by side pads to reduce side-to-side rolling motions. A neck collar is not used on the subject in Fig. 2.25, but one is shown in Fig. 2.24.

2.5.8   Sensor placement and data collection

Accelerometers, a hybrid system with accelerometers and motion capture markers, or inertia-based motion sensors can be used in the experiments. In this example, Wireless Motion Tracking Sensor (MTw)[6] were used to collect data for the experiment at 60 Hz (Fig. 2.25). Each MTw sensor measured accelerations in three directions, angular velocity in three directions, and orientation in 3D space with reference to Earth North and gravity. Synchronized data from a multitude of sensors allows the gathering of information that would not be possible otherwise.

In addition to the markers, there is another technique that can be used to reduce the effect of the magnetic field in the testing room. In a magnetically disturbed room, the local sensor yaw angle (Earth North heading) of the MTw sensor can become compromised. To reduce this effect, the experimental setup was raised 30.5 cm above the surface of the shaking platform. The raised platform was a reinforced wood structure that was tested to ensure rigidness past 25 Hz. To further reduce potential error in yaw angle, each MTw sensor was aligned with Earth North, and the heading was mathematically reset in the subsequent data processing (Frick, 2015). While the marker system was used in the experiments, its usage was limited to validate the resulting acceleration in different directions from the inertial system.

The MTw sensors used in this work weighed approximately 27 g each. An athletic tape was wrapped around the areas of the body where the sensors would be attached and was secured with duct tape, which runs over the athletic tape and sensors to reduce skin motion. Each inertial sensor was

---

4. Spine-board, North American Rescue, Greer, SC, United States.
5. Strap Restraint System, Spider Type, North American Rescue, Greer, SC, United States.
6. Xsens Technologies, Enschede, The Netherlands.

attached at each body location at the prewrapped locations using double-sided tape. The sensors were attached to measure the input motion at the transport system and the output motions at different locations of the human body. One sensor was attached at the rigid surface of the support, measuring the input motion to the system, and another sensor was attached at the interface between the human body and the supporting surfaces, on the spine board, to measure the input motion to the human body. This sensor was attached to the center of the supporting surface. Sensors were attached to the following areas on the human body: the skin of the forehead, just between the eyebrows, to measure the output motion at the head; the skin on the flattest location of the sternum, to measure the motion at the chest; the clavicle, for more accurate measurements around the neck area; the pelvic area, through a belt tightened over the left anterior superior iliac spine; and just above the patella of the left knee.

More sensors can be attached to the feet and arms, if needed, but were not used in this example. In order to reduce skin motion where relative skin-to-bone motion can be an issue, the head and the chest sensors were attached to the skin using 3.5×5 cm double-sided tape. Because relatively large skin motion is expected to occur around bony areas, the sensors on the pelvis and knee were tightly attached to the clothing of the subjects. A secure area at the pelvis was created by tightening a leather belt around the subject's pelvis, while an athletic wrap and duct tape were used to tighten the area at the knee location. The MTw wireless sensors used in the example allow data collection from a desktop or laptop computer. Commercial software[7] was used to collect and then extract the triaxial accelerometer, triaxial gyroscope, and 3D orientation data from the inertial sensors. One important issue to consider is the placement of sensors because litters can be long and flexible. In this case, the person's center of gravity may be a good choice.

2.6 Summary and concluding remarks

Measurement of human motion in response to WBV can be complicated. Although recent developments in sensor and sensing technologies have led to the invention of several sensing systems that can be used in WBV, they still have limitations. The introduction of hybrid systems in which different sensors are combined will help researchers conduct WBV experiments that simulate real-life scenarios and give them the ability to measure both translational and rotational motions. This is important because a comprehensive measurement of supine-human motion will provide more insight into what can happen to humans during WBV.

According to the guidelines in most ISO standards, such as ISO 2631-1, the measurement of the input motion to the human body is normally done at

---

7. MT Manager, Xsens Technologies, Enschede, The Netherlands.

the interface between the human body and the supporting surfaces. Because the body of the supine human during transport is distributed across the supporting surface, choosing one input location can be challenging, especially because most supporting surfaces are not rigid. Depending on the scope of the study, sensors can be attached at the interface with locations such as the head, chest, pelvis, legs, and feet. The addition of more sensors would increase cost and effort but would be useful for investigating the local performance of the supporting surfaces. In most cases, however, one sensor can be attached to the supporting surfaces close to the center of mass of the human body to measure the input motion.

Because it has long been believed that the most detrimental vibration comes from vertical input in the gravity direction, most measurements used to evaluate the performance of immobilization systems are collected in the vertical direction, and most immobilization systems are designed to mitigate vibration energy in that direction. Recent studies (Rahmatalla et al., 2020), however, have demonstrated that supine humans can be exposed to significant translational and rotational motions in the other directions. Therefore conducting measurements in six directions (up-and-down, side-to-side, fore-aft, pitch, roll, and yaw) will generate an optimal setup to evaluate transport systems subjected to any type of motion.

References

American National Standards Institute. (2002). *Mechanical vibration and shock—Evaluation of human exposure to whole body vibration—Part 1: General requirements (ANSI S3.18 2002/ISO 2631—1:1997).* Retrieved from https://webstore.ansi.org/standards/asa/ansis3182002iso26311997.

British Standards Institution. (1974). *Guide to evaluation of human exposure to whole-body vibration (BSI DD 32).* London: British Standards Institution.

Coermann, R. R. (1961). *The mechanical impedance of the human body in sitting and standing position at low frequencies (ASD Technical Report 61—492).* Aeronautical Systems Division, Air Force Systems Command, United States Air Force.

DeShaw, J., & Rahmatalla, S. (2012). Comprehensive measurement in whole-body vibration. *Journal of Low Frequency Noise, Vibration and Active Control, 31*(2), 63—74.

Dytran Instruments, Inc. (n.d.) *Triaxial VC MEMS accelerometer.* Retrieved from https://www.dytran.com/Model-7583A1-Triaxial-VC-MEMS-Accelerometer/.

Frick, E. C. (2015). *Mitigation of magnetic interference and compensation of bias drift in inertial sensors* (unpublished master's thesis). The University of Iowa.

Griffin, M. J. (1976). Eye motion during whole-body vertical vibration. *Human Factors, 18*(6), 601—606.

International Organization for Standardization. (1997). *Mechanical vibration and shock—Evaluation of human exposure to whole-body vibration—Part 1: General requirements (ISO 2631—1:1997).* Retrieved from https://www.iso.org/standard/7612.html.

Kurihara, K., Hoshino, S., Yamane, K., & Nakamura, Y. (2002). Optical motion capture system with pan-tilt camera tracking and real time data processing. In *Proceedings 2002 IEEE International Conference on Robotics and Automation (Cat. No.02CH37292).* Retrieved from https://ieeexplore.ieee.org/document/1014713.

Leardini, A., Chiari, A., Della Croce, U., & Cappozzo, A. (2005). Human movement analysis using stereophotogrammetry. Part 3. Soft tissue artifact assessment and compensation. *Gait & Posture, 21*(2), 212−225.

Magid, E. B., Coermann, R. R., & Ziegenruecker, G. H. (1960). Human tolerance to whole body sinusoidal vibration. Short-time, one-minute and three-minute studies. *Aerospace Medicine, 31*, 915−924.

Mansfield, N. (2017). Vibration and shock in vehicles: New challenges, new methods, new solutions. In *1st International Comfort Congress*. Retrieved from http://irep.ntu.ac.uk/id/eprint/ 30974/.

Meusch, J., & Rahmatalla, S. (2014). 3D transmissibility and relative transmissibility of immobilized supine humans during transportation. *Journal of Low Frequency Noise, Vibration and Active Control, 33*(2), 125−138. Available from https://doi.org/10.1260/0263-0923.33.2.125.

OptiTrack. (n.d.). *PrimeX 22*. Retrieved from https://optitrack.com/cameras/primex-22/.

Padgaonkar, A. J., Krieger, K. W., & King, A. I. (1975). Measurement of angular acceleration of a rigid body using linear accelerometers. *Journal of Applied Mechanics, 42*(3), 552−556.

Panjabi, M. M., Andersson, G. B., Jorneus, L., Hult, E., & Mattsson, L. (1986). In vivo measurements of spinal column vibrations. *The Journal of Bone and Joint Surgery. American Volume, 68*(5), 695−702.

Rahmatalla, S., DeShaw, J., Stilley, J., Denning, G., & Jennissen, C. (2018). Comparing the efficacy of methods for immobilizing the thoracic−lumbar spine. *Air Medical Journal, 37*(3), 178−185. Available from https://doi.org/10.1016/j.amj.2018.02.002.

Rahmatalla, S., DeShaw, J., Stilley, J., Denning, G., & Jennissen, C. (2019). Comparing the efficacy of methods for immobilizing the cervical spine. *Spine (Philadelphia, Pa.: 1986), 44*(1), 32−40. Available from https://doi.org/10.1097/BRS.0000000000002749.

Rahmatalla, S., Kinsler, R., Qiao, G., DeShaw, J., & Mayer, A. (2020). Effect of gender, stature, and body mass on immobilized supine human response during en route care transport. *Journal of Low Frequency Noise, Vibration & Active Control, 40*(1), 3−17.

Rahmatalla, S., Xia, T., Contratto, M., Kopp, G., Wilder, D., Frey-Law, L., & Ankrum, J. (2008). Three-dimensional motion capture protocol for seated operators in whole body vibration. *International Journal of Industrial Ergonomics, 38*, 425−433.

Rahmatalla, S., Xia, T., Contratto, M., Wilder, D., Frey-Law, L., Kopp, G., & Grosland, N. (2006). 3D displacement, velocity, and acceleration of seated operators in a whole-body vibration environment using optical motion capture systems (paper presentation). In *Ninth International Symposium on the 3-D Analysis of Human Movement*, Valenciennes, France.

Schwartz, M., & Dixon, P. C. (2018). The effect of subject measurement error on joint kinematics in the conventional gait model: Insights from the open-source pyCGM tool using high performance computing methods. *PLoS One, 13*(1), e0189984. Available from https://doi.org/10.1371/journal.pone.0189984.

Xsens. (2009). MVN user manual. Retrieved from http://www.xsens.com/en/mvn-biomech.

Chapter 3

Biodynamics of supine humans subjected to vibration and shocks

3.1 Introduction

There are many situations in which people are exposed to whole-body vibration (WBV) while they are lying face up in the supine position. Examples include sleeping in a train berth or being transported as a patient in bumpy off-road conditions en route to a hospital. Under such circumstances, the supine human body experiences both voluntary and involuntary shaking motions that can induce discomfort, pain, and injuries. The motion can come from muscle activation or intended/unintended movements. In order to reduce these motions during medical transport, patients in the supine position are normally immobilized to a transport system, which is then connected to the vehicle floor.

During transport, energy, in terms of vibration and shocks (sudden impacts), will transfer from the road through the vehicle's tires, suspension system, and floor to the transport system, then the immobilization system, and ultimately to the patient. Some of this energy can be absorbed or dissipated in the form of heat to the surrounding environment, depending on the weight of the vehicle and the mechanical and damping properties of its materials, as well as the type of padding used to support the patient. The remaining energy that is transferred to the patient can be magnified due to the resonance phenomenon seen in the transport system and the human body itself. During resonance, mechanical systems can generate motion that is larger than the motion that entered the system to begin with. As a biomechanical system, the human body has its own resonance properties and can generate large involuntary motions if the body is subjected to vibration at certain frequencies. For example, during sitting, the human body normally resonates when it is subjected to up-and-down vibration that has frequencies between 4 and 6 Hz (4−6 cycles per second) (Griffin, 1990). In other words, if we shake a person with vertical vibrations of 4−6 cycles per second, then the body will magnify the input motion and produce motion that can be two to three times larger than the original input motion. The motion of patients in the supine position can also be magnified under resonance conditions. Studies have shown that when a supine human lying on a rigid surface is

Prehospital Transport and Whole-Body Vibration. DOI: https://doi.org/10.1016/B978-0-323-90103-1.00003-6

subjected to vertical vibration in the direction perpendicular to the floor, the torso may resonate around 6–8 Hz, the pelvis may resonate around 4–6 Hz, and the head may resonate around 12–14 Hz (Meusch & Rahmatalla, 2014a, 2014b). Of course, the natural frequencies of the human body can be affected by many factors, including weight, stiffness, and body proportions.

Transmitted vibration during medical transport can affect the biodynamics of the human response in different ways, depending on how the human is immobilized, the vibration magnitude, and the frequency content of the vibration signal. Immobilization systems normally have padded surfaces to mitigate the amount of vibration transmitted to the human, and they can be complemented with straps that hold the body to the transport system to reduce motion. Vibration magnitude is normally a measure of the intensity of motion and is measured by the unit of acceleration (m/s^2, i.e., meters per second squared), while velocity is a measure of how quickly the motion changes with time (m/s). Acceleration can also be defined as a measure of how the velocity is changing with time. The frequency content of the acceleration signal is a measure of its frequency components, which can be likened to color components in natural light such as red, blue, and green. Much like the collection of color components produces white light, the collection of frequency components defines the type of vibration signal. When the ride produces slow motions with large displacements, that means it has low-frequency components, and when it produces a buzzing feeling like that produced by a drill, it has high-frequency components. Vibration is sometimes represented by shocks when the system is subjected to sudden, short-lived forces; an example is when a vehicle hits a bump in the road. The effect of shocks on the biodynamic response of a supine human can be damaging; how damaging depends on the design and materials of the transport system and how the person is strapped to it.

Other factors that can affect the biodynamic response of humans to vibration include body weight, body shape and dimension, muscle activation, and the amount of soft tissue or fat in the body. The human body is a collection of segments connected via bones, muscles, tendons, ligaments, soft tissue, blood vessels, nerve systems, organs, and fluids, among other things. Depending on its mass and its stiffness and damping (energy-dissipating) properties, each segment of the human body can react differently to vibration. For example, the skull is very stiff and can be affected differently than the chest, which comprises many soft organs and components. Because of the complexity of the human body, in vibration analysis, the human body is simplified into several masses connected by springs and dampers. Even in this simplistic form, the analysis of human response due to vibration can be very complicated. Having said that, the simplistic form of the human body can provide useful information about how it responds to vibration, which can be very helpful when designing transport and immobilization systems with human comfort and safety in mind. Most vibration measurements can now be conducted on the human body using state-of-the-art motion-sensing systems. Different types of measurement techniques are introduced in Chapter 2, Measurement of Human Response to Vibration.

This chapter is organized as follows: the following section presents the functions traditionally used to evaluate the biodynamic response of the human body during WBV. Next, the biodynamics of the human body is introduced via tests on a rigid surface with and without padding. Then the effects of straps and vibration magnitude on the human response are discussed. This is followed by a discussion of the effect of posture on the biodynamic response. The last part of the chapter will discuss the effects of gender, mass, and anthropometry on human response during transport. The chapter ends with a summary and conclusions.

3.2   Biodynamical evaluation functions

A variety of metrics have been used in the literature to evaluate the biodynamic response of the human body and the intensity of motion transferred to the human body as a result of WBV. These measures are normally mathematical functions called *transfer functions*. Transfer functions, in general, describe the relationship between the input motions or forces entering the body and the resulting output motion of the body; in other words, they describe the behavior of the body under different types of input. If the magnitude of the transfer function at a certain frequency is greater than 1, this means that the motion at the human body is magnified relative to the input motion. If the magnitude of the transfer function at a certain frequency is less than 1, then the motion at the human body is reduced. The input motion/force to the body is normally measured via motion/force sensors at the interface between the support system (such as a seat cushion in a vehicle or a stretcher in an ambulance) and the human body. The output motion is measured by attaching motion sensors directly to the body. Some popular transfer functions used in the literature to describe the biodynamic response of humans to WBV will be described below; these are transmissibility, apparent mass (AM), mechanical impedance, and absorbed power.

3.2.1   Transmissibility

The transmissibility function is widely used in practice to evaluate human response to vibration. This function represents the ratio between the input motion to the human body at the interface between the human body and the supporting surfaces and the output motion of the human body measured at different locations on the body.

Fig. 3.1 shows a schematic drawing of the human body, supporting surface, and measurement points. Transmissibility is traditionally defined as an output-only function, meaning it can be defined as a ratio between any two points.

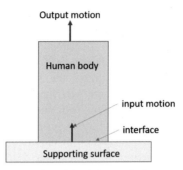

FIGURE 3.1 Schematic representation of single input–single output transmissibility.

Transmissibility is normally described as the energy through the body as the energy proceeds from the input location to the output location (Demic & Lukic, 2009).

Head-to-seat transmissibility is one well-known application of transmissibility in seated positions. In this case, the input motion is measured at the interface between the seat cushion and the person's buttocks, and the output motion is measured at the person's head. The input and output motions can be measured in terms of displacement, velocity, or acceleration; in practice, acceleration is the most used measurement. Mathematical expressions and more details about transmissibility derivations can be found in the literature (Griffin, 1990), but a simplified form is provided here:

$$T(f) = \frac{\text{output acceleration}(f)}{\text{input acceleration}(f)} \qquad (3.1)$$

Transmissibility is a function of frequency (f), which means it has a value at each frequency. In practice, the transmissibility function can take different forms depending on the number of input motions and the number of output motions, as follows:

3.2.2 Single input–single output transmissibility

In this case, the input motion has one component, and the output motion also has one component (Fig. 3.1). The expression for the transmissibility function can take the following form:

$$H_{Xx}(\omega) = S_{XX}^{-1}(\omega)S_{Xx}(\omega) \qquad (3.2)$$

where S_{Xx} represents the cross-spectral density between the input motion (X) and the output motion (x), and S_{XX} represents the autospectral density of the input motion (X).

While this type of transmissibility has a simple form, it is widely used due to its simplicity and due to historical reasons related to the limitations and difficulties of conducting and correlating measurements in multiple directions. It is the one normally used by researchers in the field because the measurement can be done with only single-axis accelerometers at the input and output locations.

Fig. 3.2 shows a generic graph that depicts the traditional trend and shape of the transmissibility function that can be seen when testing a human in a seated position. The graph normally starts with a transmissibility magnitude of 1 at lower frequencies, indicating a rigid-body motion between the input and the output. This means that the input motion will not be magnified and that the output motion will be similar to the input motion. The graph then ascends and shows a peak at the location where the resonance of the system occurs, around 5 Hz in this case. In this curve, the input motion will be magnified by more than five times if the system is exposed to an input vibration of 5 Hz. The resonance phenomenon takes place when the input frequencies match the natural frequency of the system. The natural frequency is a measure of the ratio between the stiffness of the system and its mass. Stiffer systems have a peak that is shifted to the right (at a higher frequency), and softer systems have a peak that is shifted to the left (at a lower frequency). The graph descends after the peak in most cases, and the transmissibility reaches values lower than 1 after 6 Hz. In general, multibody systems with many masses can have many natural frequencies and therefore many peaks, with each peak occurring at one of the natural frequencies.

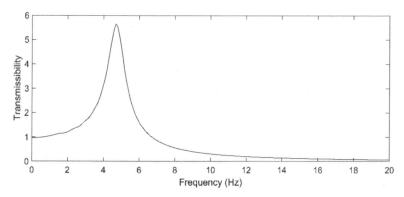

FIGURE 3.2   Schematic representation of a generic transmissibility graph.

Fig. 3.3 shows a sample transmissibility graph representing the magnitude ratio between the output vertical acceleration at the head of a seated person leaning on a seatback and the input vertical acceleration at the interface between the person and a rigid seat. The light lines represent the data from eight subjects, and the dark line represents the average of the subjects. Fig. 3.3 illustrates a clear variability between subjects (intersubject variability). It should be noted that transmissibility graphs can also differ within

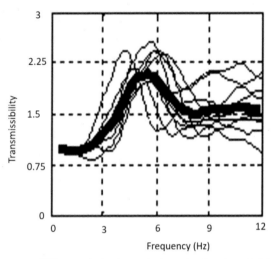

FIGURE 3.3   Transmissibility graphs in the vertical gravity direction of eight seated people with seat-back support and their average.

subjects (intrasubject variability). In general, the transmissibility graphs for this case showed peak frequencies between 4 and 6 Hz across subjects, with the average at around 5 Hz. The information from such a graph is useful in many analyses, including those for design development and ergonomics. For example, the transmissibility magnitude of 2 around 5 Hz means that if a seated person is exposed to an acceleration of 1 m/s² at 5 Hz at the seat level, the person's head will move at 2 m/s² at that frequency. At very low frequencies, the body moves with the seat as a rigid mass, and therefore its transmissibility is close to 1. It should be mentioned that transmissibility can be affected by many factors, including posture, muscle tension, and vibration magnitude.

The transmissibility function can be calculated in any direction. The following sections will demonstrate scenarios where the transmissibility can have many components for different directions. For more information on transmissibility, see the literature (Paddan & Griffin, 1998; Rahmatalla & DeShaw, 2011; Smith et al., 2008).

3.2.3   3D-multiple input−3D-multiple output transmissibility

This type of transmissibility has input and output in the three translational directions. Triaxial sensors (accelerometers) should be attached at the interface (input) between the contact surface and the human body, as well as at points on the human body such as the head (output). Fig. 3.4 shows a schematic representation of this type of setup.

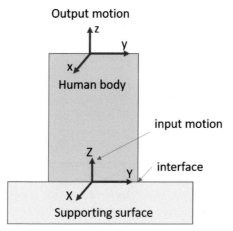

FIGURE 3.4   Schematic representation of 3 input−3 output transmissibility.

The mathematical form of the resulting 3D-multiple input−3D-multiple output (3IP−3OP) transmissibility is as follows (Newland, 1984):

$$\begin{bmatrix} H_{Xx} & H_{Xy} & H_{Xz} \\ H_{Yx} & H_{Yy} & H_{Yz} \\ H_{Zx} & H_{Zy} & H_{Zz} \end{bmatrix} = \begin{bmatrix} S_{XX} & S_{XY} & S_{XZ} \\ S_{YX} & S_{YY} & S_{YZ} \\ S_{ZX} & S_{ZY} & S_{ZZ} \end{bmatrix}^{-1} \begin{bmatrix} S_{Xx} & S_{Xy} & S_{Xz} \\ S_{Yx} & S_{Yy} & S_{Yz} \\ S_{Zx} & S_{Zy} & S_{Zz} \end{bmatrix} \qquad (3.3)$$

where H_{Xx} represents the transmissibility component between the input motion in direction X and the output motion in direction x; S_{Xx} represents the cross-spectral density between the input motion in the X direction and the output motion in the x-direction; and S_{XX} represents the autospectral density of the input motion in the X-direction. As shown in Eq. (3.3), this type of transmissibility has nine components, which makes it much harder to use than single input−single output (SIP−SOP) for subsequent analyses and designs. For example, in the design of a seat or litter system for transport, this type of transmissibility will result in nine graphs. Three of them, the diagonal components (H_{Xx}, H_{Yy}, and H_{Zz}), will have strong correlations between the input and output motions in these directions because the input and output motions are in the same direction to each other. Meanwhile, the six out-of-diagonal (coupled) components, in which the direction of the output motion is orthogonal to the direction of the input motion, may present weaker correlations between the input and output motions. They normally show lower coherence, which is an indication of a nonlinear relationship. One way to simplify the analysis of multiaxis transmissibility, as done by some researchers (Mandapuram, 2012; Mandapuram et al., 2010), is to combine the transmissibility response components to a single axis vibration.

3.2.4 Single-input–3D-multiple output transmissibility

The vibration in the vertical gravity direction is normally considered the main dominant vibration component that researchers need to deal with. This could be true for many applications. Still, vibration in the fore-aft and side-to-side directions can also be dominant in some machines, such as those used in mining industries. Therefore analysis with a single-input vibration is widely used in the literature. However, the interest on the output side could be in the multiple directions. In single-input–3D-multiple output transmissibility (SIP–3OP), a uniaxial sensor is installed at the interface between the person and the contact surfaces (input), while a triaxial accelerometer or six-degree inertial sensor is attached to the human body (output), as shown in Fig. 3.5. Due to the difficulties of aligning the triaxial accelerometer on the human body so that its axes are aligned with the directions of the input sensor or in the global lab directions, some researchers derived formulations to relate the seat-to-head transmissibility response of a dual-axis horizontal vibration to the responses of a single-axis vibration (Mandapuram et al., 2010).

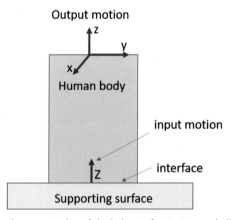

FIGURE 3.5   Schematic representation of single-input–3 output transmissibility.

It is worth mentioning here that the results of this SIP–3OP type of transmissibility will be affected by the location of the triaxial sensor on the human body. If the output sensor is attached to the back of a seated person [e.g., at the tenth thoracic vertebra (T10)] to study the motion of the lower back area, and if the input motion is in the vertical direction, then most of the motion at the output will be in the vertical direction (in the direction of the input motion). The motion in the other lateral directions, fore-aft and side-to-side, will be minimal. However, if the output sensor is attached to the head of a seated person, the pitching motion (like nodding yes) of the head will result in the vertical as well as the fore-aft components contributing to the output motion, and not much activity will be seen in the side-to-side direction. In this case, the direction of the output motion will change with time during the experiments, and its effect should be considered in the analysis.

The mathematical expression for this type of transmissibility can be derived as a subset of the 3IP−3OP transmissibility function and can be expressed as follows:

$$
\begin{bmatrix}
H_{Zx} & H_{Zy} & H_{Zz} \\
0 & 0 & 0 \\
0 & 0 & 0
\end{bmatrix}
\tag{3.4}
$$

As shown in Eq. (3.4), this form of transmissibility matrix is similar to 3IP−3OP but has only three components in one of its rows.

3.2.5 6D-multiple input−6D-multiple output transmissibility

This is the most general case, where three translational and three rotational motion components are considered at the interface between the human body and the contact surfaces (input) and at a point on the human body (output), as shown in Fig. 3.6. Inertial sensors would be a good candidate for measuring the motion at the input and output for such applications.

Eq. (3.2) can be expanded in the same way for 6D-multiple input−6D-multiple output (6IP−6OP) vibration using three translational and three rotational inputs and three translational and three rotational outputs. The

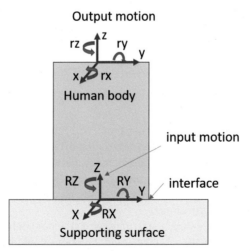

FIGURE 3.6 Schematic representation of 6 input (3 translation and 3 rotation)−6 output (6IP−6OP) transmissibility.

resulting full 6×6 transmissibility matrix is shown in Eq. (3.5) (DeShaw & Rahmatalla, 2014a; Newland, 1984).

$$
\begin{bmatrix}
H_{Xx} & H_{Xy} & H_{Xz} & H_{Xrx} & H_{Xry} & H_{Xrz} \\
H_{Yx} & H_{Yy} & H_{Yz} & H_{Yrx} & H_{Yry} & H_{Yrz} \\
H_{Zx} & H_{Zy} & H_{Zz} & H_{Zrx} & H_{Zry} & H_{Zrz} \\
H_{RXx} & H_{RXy} & H_{RXz} & H_{RXrx} & H_{RXry} & H_{RXrz} \\
H_{RYx} & H_{RYy} & H_{RYz} & H_{RYrx} & H_{RYry} & H_{RYrz} \\
H_{RZx} & H_{RZy} & H_{RZz} & H_{RZrx} & H_{RZry} & H_{RZrz}
\end{bmatrix}
$$

$$
=
\begin{bmatrix}
S_{XX} & S_{XY} & S_{XZ} & S_{XX} & S_{XY} & S_{XZ} \\
S_{YX} & S_{YY} & S_{YZ} & S_{YX} & S_{YY} & S_{YZ} \\
S_{ZX} & S_{ZY} & S_{ZZ} & S_{ZX} & S_{ZY} & S_{ZZ} \\
S_{XX} & S_{XY} & S_{XZ} & S_{XX} & S_{XY} & S_{XZ} \\
S_{YX} & S_{YY} & S_{YZ} & S_{YX} & S_{YY} & S_{YZ} \\
S_{ZX} & S_{ZY} & S_{ZZ} & S_{ZX} & S_{ZY} & S_{ZZ}
\end{bmatrix}^{-1}
\begin{bmatrix}
S_{Xx} & S_{Xy} & S_{Xz} & S_{Xx} & S_{Xy} & S_{Xz} \\
S_{Yx} & S_{Yy} & S_{Yz} & S_{Yx} & S_{Yy} & S_{Yz} \\
S_{Zx} & S_{Zy} & S_{Zz} & S_{Zx} & S_{Zy} & S_{Zz} \\
S_{Xx} & S_{Xy} & S_{Xz} & S_{Xx} & S_{Xy} & S_{Xz} \\
S_{Yx} & S_{Yy} & S_{Yz} & S_{Yx} & S_{Yy} & S_{Yz} \\
S_{Zx} & S_{Zy} & S_{Zz} & S_{Zx} & S_{Zy} & S_{Zz}
\end{bmatrix}
$$

$$(3.5)$$

The resulting transmissibility matrix has 36 transmissibility components, that is, 36 transmissibility graphs that need to be dealt with. Some of these transmissibilities are related to the relationship between the translational components and some to the relationship between the rotational components. In addition, some of the transmissibility components are related to rotation input motion with translational output motions and vice versa. One way to simplify this matrix is by partitioning it into four quadrants as shown in the following form:

$$
\left[
\begin{array}{ccc:ccc}
H_{Xx} & H_{Xy} & H_{Xz} & H_{Xrx} & H_{Xry} & H_{Xrz} \\
H_{Yx} & H_{Yy} & H_{Yz} & H_{Yrx} & H_{Yry} & H_{Yrz} \\
H_{Zx} & H_{Zy} & H_{Zz} & H_{Zrx} & H_{Zry} & H_{Zrz} \\
\hdashline
H_{RXx} & H_{RXy} & H_{RXz} & H_{RXrx} & H_{RXry} & H_{RXrz} \\
H_{RYx} & H_{RYy} & H_{RYz} & H_{RYrx} & H_{RYry} & H_{RYrz} \\
H_{RZx} & H_{RZy} & H_{RZz} & H_{RZrx} & H_{RZry} & H_{RZrz}
\end{array}
\right]
$$

$$(3.6)$$

In this case, the upper left quadrant will comprise the relationship between the translational input and translational output motions, the lower right quadrant will comprise the relationship between the rotational input and the rotational output motions, and the lower left and upper right quadrants will comprise the relationship between the mixed translational and rotational input and output directions.

While the pure three-directional translational input−output motions and three-directional pure rotational input−output motions could be simplified and transformed to a SIP−SOP form by taking the norm of the input and output motions, it is very hard to do so with the 6IP−6OP transmissibility. One reason for the difficulty is the differences between the units of the translational and rotational components. If inertial sensors are used, acceleration is normally expressed in m/s^2 for the translational components, and angular velocity is expressed in rad/s for the rotational components. Also, the out-of-diagonal components will have units that are a mixture of acceleration and angular velocity. One way to circumvent these issues is to use the effective transmissibility, where it is possible to combine the 36 transmissibility components into four graphs representing the translational transmissibility term, the rotational transmissibility terms, and the mixed translational−rotational terms.

3.2.6   Effective transmissibility

While transmissibility is widely used in the literature because of its simplicity in terms of measurement and interpretation, the increase in the number of motion components when dealing with multiple inputs and multiple outputs can cause the calculation and interpretation to become more involved. In their work on the transmissibility of vibration to the car seatback, Qiu and Griffin (2004) showed that the SIP−SOP model can reasonably be used to study the transmissibility to the seatback with multiple input vibrations in the vertical direction but was less effective for the fore-aft direction. While the calculation can be easily handled by existing commercial software like MATLAB, interpreting the results from multidirectional tests is challenging. It requires dealing with many graphs, with each graph telling a part of the story, as well as with the physical meaning of the transmissibility components. One issue is how the biodynamic response of humans is affected by the direct (diagonal) components in the transmissibility matrix; another is the role of out-of-diagonal components in the transmissibility matrix and their effect on the biodynamic response of humans. So the question is whether it is possible to combine the 36 components in the transmissibility function and transform them into one graph, similar to that of the SIP−SOP transmissibility, which researchers are familiar with and know how to interpret. The answer is yes, in part, and some researchers are already doing this using different approaches. One approach mentioned earlier in the section on 3IP−3OP transmissibility is to transform the transmissibility response of a dual-axis horizontal vibration to the responses of a single-axis vibration. While this simplified form can be effective, the implementation of this approach for 6IP−6OP can be challenging.

Another way to combine the components of the transmissibility function for multiple inputs and outputs is to use suitable mathematical expressions such as singular value decomposition (SVD) (DeShaw & Rahmatalla, 2014a; Rahmatalla & DeShaw, 2011). SVD is a mathematical method and is an effective way to extract the principal components of a rectangular matrix with their principal directions. The principal components can be considered the eigenvalues of the matrix, and their directions can be considered the associated eigenvectors (Heath, 1997).

For a matrix like **H** with $m \times n$ elements, the SVD process will produce three matrices as follows:

$$SVD \left\{ \begin{bmatrix} H_{Xx}(\omega) & H_{Xy}(\omega) & H_{Xz}(\omega) \\ H_{Yx}(\omega) & H_{Yy}(\omega) & H_{Yz}(\omega) \\ H_{Zx}(\omega) & H_{Zy}(\omega) & H_{Zz}(\omega) \end{bmatrix} \right\} = \mathbf{U}\mathbf{\Sigma}\mathbf{V}^T \qquad (3.7)$$

where **V** and **U** are unitary matrices, that is, $\mathbf{V}^*\mathbf{V} = \mathbf{V}\mathbf{V}^* = \mathbf{I}$; **I** is the identity matrix; $\mathbf{V}^*$ is the complex conjugate of **V**; and $\mathbf{\Sigma}$ is a diagonal matrix having the following form:

$$\mathbf{\Sigma}(\omega) = \begin{bmatrix} H_{11}(\omega) & 0 & 0 \\ 0 & H_{22}(\omega) & 0 \\ 0 & 0 & H_{33}(\omega) \end{bmatrix} \qquad (3.8)$$

As shown in Eq. (3.8), $\mathbf{\Sigma}$ is a function of (ω) or the frequency. With this new matrix form, it becomes much easier to deal with only the diagonal components. This new form can be transformed to a single number at each frequency by simply combining them using the mathematical expressions of the norm approach. The results of the latter process will produce an expression with a graph that is analog to the SIP−SOP transmissibility (DeShaw & Rahmatalla, 2014a; Rahmatalla & DeShaw, 2011).

Another mathematical approach with more physical meaning can also be used to transfer the $\mathbf{\Sigma}$ matrix to a single value at each frequency. This approach is based on the idea that the transmissibility function represents the energy through the body, as it correlates the input energy entering the body to the output energy on the body (Hinz & Seidel, 1987). Based on this concept, the transmissibility through the body can be considered a stress-like quantity, where the principal of the maximum distortion energy theory (Hibbeler, 2008) can be implemented to determine a representative component or a resultant number at each frequency. The resulting function of frequencies can be considered to represent the SIP−SOP form of the transmissibility, which can be called the effective transmissibility (H_{eff}) (DeShaw & Rahmatalla, 2014a; Rahmatalla & DeShaw, 2011), where

$$H_{\text{eff}} = \sqrt{H_{11}^2 + H_{22}^2 + H_{33}^2 - H_{11}H_{22} - H_{22}H_{33} - H_{11}H_{33}} \qquad (3.9)$$

The generalization of Eq. (3.9) to 6IP−6OP is straightforward, but there is a problem with the units as indicated above. The units of the first quadrant in the 6IP−6OP matrix in Eq. (3.6) are normally acceleration, and the units in the fourth quadrant are normally angular velocity. One way to circumvent this problem is to take the SVD for each quadrant. This process will produce four submatrices, each of which can be transferred to SIP−SOP transmissibilities using Eq. (3.9). One of them, the upper left quadrant, represents the translational biodynamic response to the translational input motions; another, the lower right quadrant, represents the rotational biodynamic response to the rotational input motions; and the remaining two quadrants represent the mixing biodynamic response, that is, the translational response to the rotational input and the rotational response to the translational input.

3.2.7   Case study: effective transmissibility

3.2.7.1   Single input–3D multiple output condition

This case study demonstrates the use of the effective transmissibility in evaluating the effect of posture on the biodynamic response of 12 seated humans under different scenarios of WBV in multiple directions (DeShaw & Rahmatalla, 2014a). The subjects experienced a single-input random vibration in the vertical direction, and measurements were conducted on the subjects' heads in the three translational directions (x, y, z). Fig. 3.7 shows the

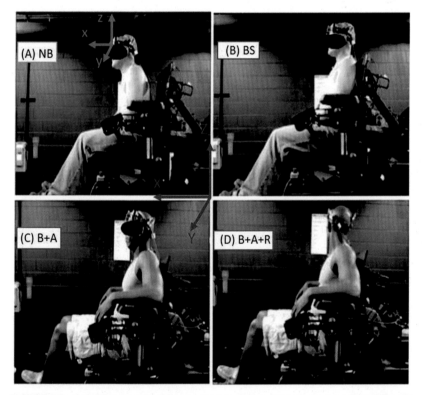

FIGURE 3.7   Seated humans with different posture conditions: (A) no-backrest condition (NB), (B) backrest-supported condition (BS), (C) backrest-supported and armrest-supported condition (B + A), and (D) backrest-supported, armrest-supported, and head-rotated condition (B + A + R). *Adapted from DeShaw J., & Rahmatalla, S. (2014a). Effective seat-to-head transmissibility under combined-axis vibration and multiple postures.* International Journal of Vehicle Performance, *1(3/4), 235. https://doi.org/10.1504/IJVP.2014.069108.*

experimental setups for the different posture conditions. In Fig. 3.7A, the subject leans his back away from the seatback; this condition is called no-backrest (NB). In Fig. 3.7B, the subject leans his back on the seatback; this condition is called backrest-supported (BS). In Fig. 3.7C, the subject leans his back on the seatback and puts his arms on the armrests, so this condition is called backrest-supported and armrest-supported (B + A). Finally, the subject in Fig. 3.7D leans his back on the seatback, puts his arms on the armrests, and twists his neck by 90 degrees and looks to the side; this condition is called backrest-supported, armrest-supported, and head-rotated (B + A + R).

Fig. 3.8 shows the resulting transmissibility for one condition (BS). As shown in Fig. 3.8A, the transmissibility functions for this condition have three components represented by three graphs correlating the input (X, Y, and Z) and output (x, y, and z) motions. Fig. 3.8B shows the effective transmissibility of this condition, where one graph combines the information in Fig. 3.8A. The effective transmissibility shows a large peak around 5 Hz. This peak matches well with that of the vertical-to-vertical (Z_z) transmissibility component, which played a major role under the conditions of this test; this could have also happened because the main input energy was in the vertical Z direction.

Using the concept of effective transmissibility, the biodynamic response of the subjects under three different inputs (vertical [Z_z], fore-aft [X_x], and side-to-side [Y_y]) and the four posture conditions are shown in Fig. 3.9.

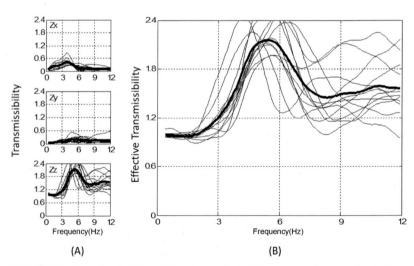

(A)   (B)

FIGURE 3.8 Single input−3D-multiple output transmissibility components under the backrest-supported posture: (A) three graphs representing the transmissibility in the Z_x, Z_y, and Z_z directions; (B) one graph representing the effective transmissibility of 12 participants, with the bold line representing the average of the group. *Adapted from DeShaw J., & Rahmatalla, S. (2014a). Effective seat-to-head transmissibility under combined-axis vibration and multiple postures.* International Journal of Vehicle Performance, *1(3/4), 235. https://doi.org/10.1504/IJVP.2014.069108.*

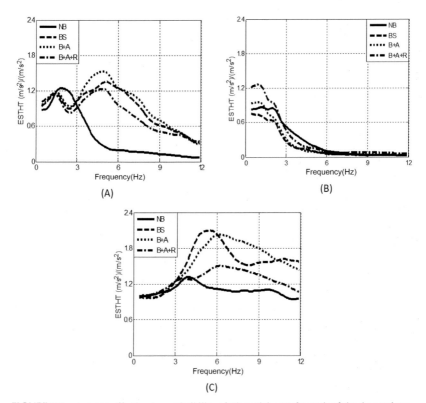

FIGURE 3.9 Average effective transmissibility of 12 participants for each of the 4 seated postures during (A) random fore-aft vibration (X_x), (B) random lateral vibration (Y_y), and (C) random vertical vibration (Z_z). *Adapted from DeShaw J., & Rahmatalla, S. (2014a). Effective seat-to-head transmissibility under combined-axis vibration and multiple postures.* International Journal of Vehicle Performance, 1(3/4), 235. https://doi.org/10.1504/IJVP.2014.069108.

Fig. 3.9A shows that the BS and B + A postures had the highest vibration transfer to the subject's head during fore-aft input vibration. The NB posture shifted the transmissibility graphs (X_x) from around 5 to 2 Hz, indicating a stiffening effect to the head−neck region. When the input vibration was in the lateral side-to-side direction (Y_y), shown in Fig. 3.9B, the B + A + R condition showed the highest peak transmissibility. This could possibly be attributed to the pitching motion of the head, which produces a lateral movement when the head is twisted to the side. In addition, the transmissibility graphs show that the armrest had little effect on the vibration transferring to the head. Fig. 3.9C shows the situation under the vertical input vibration. In this case, the BS condition showed the highest peak transmissibility. This could be related to the effect of back support, which gives support and stabilizes the upper torso and allows most of the input energy to transfer to the head and cause more head motion. This could also explain why the head motion in the NB posture was lower than in the supported postures. The smaller peak in the NB posture (Fig. 3.9C) could be due to the coupling between the head/neck and the upper torso muscles, which will combine and act to balance the upper body during vibration and can result in a stiffening of the cervical spine area. The effect of neck stiffening during the NB posture can be also seen in Fig. 3.9A where the peak frequency was shifted to the left to a lower value than in the other posture conditions.

3.2.7.2 3D-multiple input−3D-multiple output condition

Under this condition, the seated human subjects were exposed to three translational random vibrational input motions, and the measurement on their heads (output motion) had three translational components. The resulting transmissibility function has nine graphs. Fig. 3.10A shows the nine graphs for the B + A + R posture. The out-of-diagonal components such as X_z, Z_x, and Y_z show more activity under this condition, some of which may be attributed to the higher input energy in these directions and to the twisted head posture. The mean effective transmissibility graph in Fig. 3.10B shows a major peak around 5 Hz, which could reflect the denominate behaviors seen in the Z-direction in Fig. 3.10A. As can be seen in this example that the analysts can not only use the information from the effective transmissibility graph in their design and ergonomics assessments but can also go back and forth between Fig. 3.10A and B to investigate the system's behaviors in different directions.

Fig. 3.11 shows the effective transmissibility for the different postures in Fig. 3.10; the BS condition had the largest transmissibility peak, and the NB

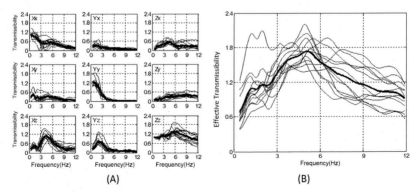

FIGURE 3.10 3D-multiple input−3D-multiple output transmissibility graphs under the backrest-supported, armrest-supported, and head-rotated posture: (A) nine transmissibility components; (B) effective transmissibility of the nine components in (A), with the bold line representing the average of 12 subjects. *Adapted from DeShaw J., & Rahmatalla, S. (2014a). Effective seat-to-head transmissibility under combined-axis vibration and multiple postures.* International Journal of Vehicle Performance, 1(3/4), 235. https://doi.org/10.1504/IJVP.2014.069108.

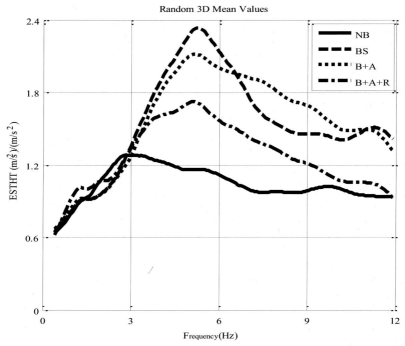

FIGURE 3.11   The average effective transmissibility graphs of 12 seated subjects for the 4 postures under the 3D-multiple input−3D-multiple output vibration condition. *Adapted from DeShaw J., & Rahmatalla, S. (2014a). Effective seat-to-head transmissibility under combined-axis vibration and multiple postures.* International Journal of Vehicle Performance, *1(3/4), 235. https://doi.org/10.1504/IJVP.2014.069108.*

condition had the lowest value. This behavior was also seen in the SIP−3OP vibration condition. The NB graph also shows the stiffest neck condition, under which smaller head motion would take place. While this could represent the best condition, the muscle activation required with this type of posture may induce fatigue with time. In fact, a study by DeShaw and Rahmatalla (2016b) showed that the use of a lumbar support can present one effective way to provide the needed back support and comfort to seated people and, at the same time, can reduce the head motion that can result when the upper torso becomes in contact with the seatback, as with the BS posture.

3.2.7.3   6D-multiple input−6D-multiple output condition

Under this condition, the seated humans were exposed to six types of motion composed of three translational inputs and three rotational inputs, representing the most general input condition that may reflect real-life field conditions. Fig. 3.12 shows that the transmissibility function for the NB posture,

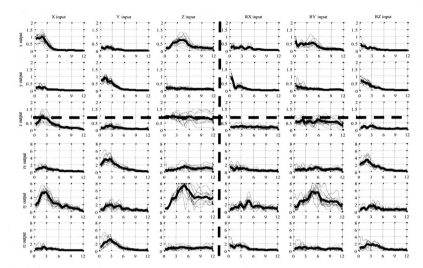

FIGURE 3.12 Transmissibility graphs and their average for 12 subjects in the no-backrest posture under the 6D-multiple input−6D-multiple output random vibration condition. *Adapted from DeShaw J., & Rahmatalla, S. (2014a). Effective seat-to-head transmissibility under combined-axis vibration and multiple postures.* International Journal of Vehicle Performance, 1*(3/4), 235.* https://doi.org/10.1504/IJVP.2014.069108.

as an example, has 36 graphs. The activities in the diagonal and out-of-diagonal matrices in each quadrant can be very challenging for analysts. However, by carefully inspecting the 36 graphs, one can find each quadrant containing nine transmissibility components that share the same units. The upper-left quadrant comprises the translational units (m/s^2) input−output transmissibility components; the lower right-quadrant comprises the rotational units (rad/s) input−output transmissibility components; and the upper-left and lower-right quadrants contain the mixed units (m/s^2 and rad/s) translational−rotational components. In this case, the concept of the effective transmissibility can be applied to each quadrant that shares the same units so that each quadrant becomes one graph as can be seen in Fig. 3.13.

Fig. 3.13 shows the effective transmissibility graphs of the four quadrants in Fig. 3.12 for the four posture conditions (NB, BS, B + A, and B + A + R) under 6IP−6OP. As can be seen from Fig. 3.13, the effective transmissibility graphs in the translational input−output directions (Fig. 3.13A) and mixed translational input and rotational output directions (Fig. 3.13C) show more activities and clear trends between the different posture conditions, as compared to the transmissibility graphs with the rotational input−output directions (Fig. 3.13B) and rotational input and translational output directions (Fig. 3.13D). These behaviors could result because of the higher energy of the input motion in the translational directions, used in this example, as compared to the input energy in the rotational directions. It should be mentioned

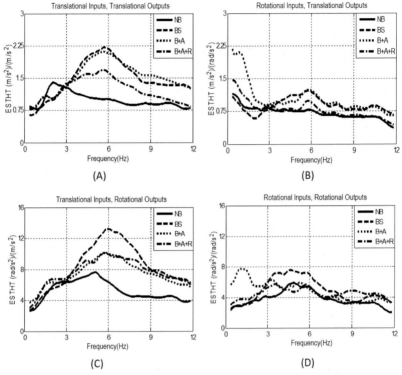

FIGURE 3.13 Average effective transmissibility graphs of the no-backrest, backrest-supported, backrest-supported and armrest-supported, and backrest-supported, armrest-supported, and head-rotated under 6D-multiple input−6D-multiple output random vibration: (A) effective 3D-multiple input−3D-multiple output (3IP−3OP) translational components, (B) effective 3IP−3OP rotational components, (C) effective 3IP−3OP translational input and rotational output components, and (D) effective 3IP−3OP rotational input and translational output components. *Adapted from DeShaw J., & Rahmatalla, S. (2014a). Effective seat-to-head transmissibility under combined-axis vibration and multiple postures.* International Journal of Vehicle Performance, 1(3/4), 235. https://doi.org/10.1504/IJVP.2014.069108.

here that the effective transmissibility graphs in Fig. 3.13 are similar to those of 1D input−6D-multiple output presented in the literature (Paddan & Griffin, 1988, 1998).

3.2.8   Apparent mass

AM is another popular transfer function and has been explored mostly in seated positions. It represents the ratio between the input force measured at the interface between the human body and the supporting surfaces and the resulting input acceleration measured at the same location (Mansfield, 2005a).

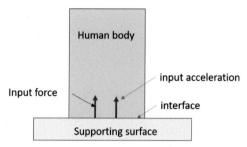

FIGURE 3.14 Schematic representation of apparent mass.

Fig. 3.14 provides a schematic representation of the measurement and locations required to calculate the AM. More details about the definition and application of AM can be found in the literature (Mandapuram et al., 2010; Mansfield et al., 2001). AM is another popular transfer (Mansfield, 2005a). Fig. 3.14 provides a schematic representation of the measurement and locations required to calculate the AM.

Peaks in AM indicate the frequencies where resonance takes place. Normally, the first prominent peak is of most interest. AM is traditionally used for the seated position, where the person's body mass is mostly supported by the seat pan. Because it provides global information about body biodynamics, it is not specific to a certain segment. There are also difficulties in identifying the locations where the input forces should be measured, especially when the supporting surfaces are flexible or cushioned (Liu & Qiu, 2020).

A simplified mathematical expression of AM can be expressed as follows:

$$AM(f) = \frac{\text{Input force } (f)}{\text{input acceleration } (f)} \qquad (3.10)$$

Fig. 3.15 provides a schematic representation of a generic AM graph of a sitting person. At very low frequency, the apparent mass shows the value of the static mass of the body, 50 kg in this case. As the frequency increases, the magnitude of the AM is affected by the motion and inertia of the body rather than by gravity alone. The graph of the AM ascends until it reaches the peak resonance frequency of the system, around 5.5 Hz in this case. After the peak frequency, the graph descends to lower values at higher frequencies. In practice, the graph of the AM could show smaller peaks in the higher frequencies after the main peak.

As with transmissibility, AM can be defined for different directions. It should be mentioned that the calculation of AM can be challenging because force measurements can be affected by how the person interacts with the equipment. For example, in seated positions, measurements will differ if the person's feet are hanging or if they are touching a foot pedal that supports some of the weight. One of the most difficult tasks is measuring the AM on

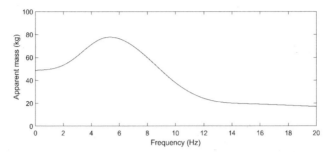

FIGURE 3.15 Schematic representation of a generic apparent mass graph.

a cushioned surface (Huang, 2008; Liu & Qiu, 2020) because the force measurements can change from location to location. One way to get around this problem is by taking measurements at different locations and then using the resultant values in the calculation. AM has been used by many researchers to investigate the biodynamics of human response to vibration in a variety of situations (Mansfield, 2005b).

3.2.9 Driving point mechanical impedance

Mechanical impedance is another popular measure of human response to vibration. It is defined as the ratio of the input force to the input velocity at the point where the force is applied (Fig. 3.16). Both the force and velocity are measured at the interface between the human body and the supporting surfaces. The velocity is normally determined by integrating the acceleration signal in the time domain. A simplified expression of impedance is as follows:

$$\text{IM}\,(f) = \frac{\text{Input force }(f)}{\text{input velocity }(f)} \tag{3.11}$$

Fig. 3.17 presents a generic graph of the impedance function of a seated person. The graph shows characteristics similar to those of other transfer functions in terms of a peak value at the resonance region. Impedance was used early on in aerospace operations to investigate the effects of vibration and impact on human response to vibration under different postures and padding conditions (Coermann, 1962). Mechanical impedance was also used by Vogt et al. (1973) in their investigation of supine human response to sustained acceleration. More details about the impedance function in the study of human response to vibration can be found in the work of Mansfield (2005b).

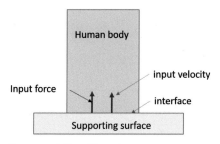

FIGURE 3.16 Schematic representation of impedance.

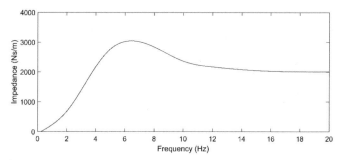

FIGURE 3.17 Schematic representation of a generic impedance graph.

3.2.10 Absorbed power

Absorbed power (AP) is another transfer function that is widely used in the literature (Holmlund et al., 2000; Mansfield et al., 2001). Researchers have used it in various applications and demonstrated the differences between it and other transfer functions, such as AM. It has been used in many studies on human response to vibration and has been very useful in comparing environments such as different types of vehicles. It is also considered a good measure for assessing the risk of injury.

AP represents the instantaneous power transmitted to the system. Mathematically, it is defined as the product of the input force $F(f)$ and input velocity $v(f)$ at the interference between the human body and the supporting surfaces (see Fig. 3.18).

$$AP(f) = \text{Input force}(f) \times \text{input velocity}(f) \qquad (3.12)$$

The velocity (v) is normally calculated by integrating the acceleration signal, and the force is normally proportional to the mass of the person, which will also be affected by vibration. Therefore the AP will sense the changes in the mass of the person and will be proportional to the vibration magnitude.

Fig. 3.19 is a schematic representation of a generic AP graph. Again, the graph depicts a peak where the system starts resonating or showing extreme

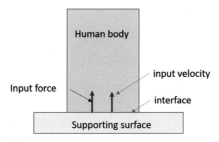

FIGURE 3.18   Schematic representation of absorbed power.

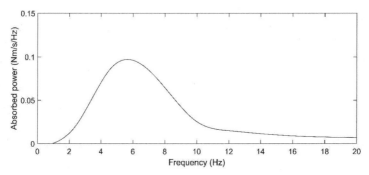

FIGURE 3.19   Schematic representation of a generic absorbed power graph.

motion. As with the transmissibility, the expression of the AP in multiple directions can be derived and used (Mandapuram et al., 2015).

3.3   Experimentation in supine transport

The transmissibility function will be used in the implementation and discussion of the examples in this chapter. As stated earlier, the input motion of the transfer functions should be measured at the interface of the human body and the supporting surfaces. Therefore it is expected that the mechanical properties of the supporting surfaces will play a major role in how the energy will be transferred to the human body. Fig. 3.20 is a schematic representation of supine-human transport in a vehicle. It shows that vibration energy can flow from the tires to the suspension system, the vehicle floor, and the transport system before entering the human body. Therefore the transport system can play a major role in controlling the energy entering the human body. If strapping is involved in the process, then the straps are also expected to play a role in the human biodynamic response. As shown in Fig. 3.20, most transport systems are designed to absorb the vibration energy in the vertical gravity direction and cannot do much to dissipate the energy in the other lateral directions. For example, if the vehicle makes a sudden braking motion, the human may slide on the surface of the transport system in the fore-aft

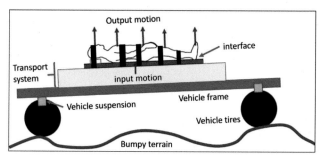

FIGURE 3.20   Schematic of transport system.

direction even if strapped to the transport system, which may involve some unwanted consequences.

Determining the biodynamic response of a human can be a straightforward process, but caution should be taken because things can go wrong during the experimental setup, measurements, or data processing and analysis. A general description of the biodynamics of the human body is provided here; more details are provided in future chapters.

Unlike seated transport, during supine-human transport, the human body will contact the supporting surfaces of the transport system at several locations, including the head, torso, pelvis, legs, and feet (Fig. 3.20). Transmissibility measurements will differ for each segment, so multiple graphs may be necessary. For example, transmissibility measurements of the head will be useful when investigating cushion materials beneath the head. While each segment can be moved locally, the segments of the human body are connected, and their motions can be affected by the motion of adjacent segments and by the motion of the body as a whole. For example, the relative motion between the head and the torso provides critical information about what is happening at the neck. If a neck collar is used, then motion information at the neck will indicate how it is affecting/restricting the relative motion between the head and torso. The motion ratio between adjacent segments is known as relative transmissibility (Meusch & Rahmatalla, 2014a).

Traditionally, most of the work on evaluating the biodynamics of humans in seated, standing, and supine positions has focused on motion in the vertical gravity direction; less work has been done in the lateral (fore-aft and side-to-side) directions. In fact, most vibration mitigation systems are implemented in the vertical direction as it has traditionally been considered the main cause of back injuries during sitting. This may be attributed to limitations in the sensing systems that were available when international standards, such as those by the International Organization for Standardization (1997), were written. More recent studies (Mansfield & Maeda, 2007) have shown the importance of considering the lateral directions while evaluating the biodynamic response of humans, and some changes are being seen on the market. For example, there are new seating systems equipped with vibration mitigation systems in the vertical and lateral directions. It is hoped that similar developments will evolve in the area of supine-human transport.

While the effects of vibration in the translational directions have been widely investigated when it comes to human response, little work has been done to investigate the effects of rotational vibration such as a vehicle rocking from side to side or front to back (Paddan & Griffin, 1994). Recent studies (Rahmatalla et al., 2020), however, have shown that rotational motions can play a considerable role in supine-human biodynamics and that these motions can be affected by gender, body mass, and anthropometry.

3.3.1   Experimental setup

During supine transport, people can be exposed to different types of vibration rides, including sine wave form functions (such as those when a person goes repeatedly up and down with the same motion), random vibration (when the motion is changing in an unpredictable manner), and shocks (when a sudden jump in motion occurs). In order to simulate real-life situations in the lab, data must first be collected from transport vehicles in the field using a human subject. The data can then be simulated in the lab setting using a shaker table. A shaker table is basically a machine that can reproduce motion in different directions. Different hypothetical types of ride files can be generated by a computer to simulate general conditions that may be captured from the field data, such as the inclusion of random signals with shocks. Different rides can also be generated on the shaker table to investigate certain phenomena at certain frequencies. Advanced shaker tables can move in the three translational directions and the three rotational directions, giving them the ability to simulate most of the vibration and shocks that can occur during prehospital transport. After obtaining the required data, the human subject can be tested on the shaker table using different configurations. For example, the human can be tested using different existing immobilization systems with various padding conditions to investigate their effectiveness. The human can also be tested on a rigid surface to investigate the biodynamic response of the human body in isolation from the transport systems. Fig. 3.21 shows an example of vibration and vibration-with-shock ride files. In the vibration-with-shock files, spikes occur for very short durations and at much higher magnitudes than the normal fluctuations that occur during vibration.

Fig. 3.22 shows a supine human on a shaker table. The subject is lying on a spinal board that is rigidly attached to the platform of the shaker table. Straps like those used by the United States Army Aeromedical Research Lab (USAARL) secure the subject. Medics normally control the tension of the straps by allowing two- to three-finger widths between the straps and the body, which allows the patient to breathe comfortably. In lab testing, the tension can be controlled and measured; in this example, it is set at approximately 135 N. Motion sensors are attached to the head, chest, pelvis, thigh, and shin.

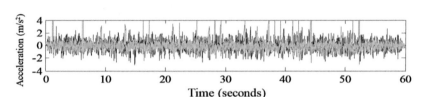

FIGURE 3.21 Time history profiles of 3D vibration (*black* lines) and 3D vibration with shocks (*green* lines); a root-mean-square magnitude around 0.8 m/s^2 was used in each of the three directions. *Adapted from Rahmatalla, S., DeShaw, J., & Barazanji, K. (2017). Biodynamics of supine humans and interaction with transport systems during vibration and shocks.* Journal of Low Frequency Noise, Vibration and Active Control, 38(2), 1−9.

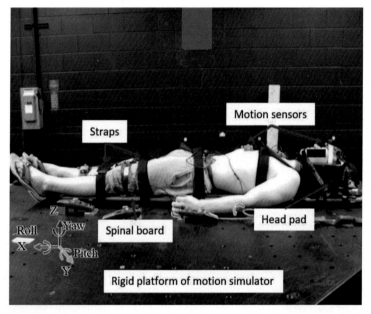

FIGURE 3.22 Supine human strapped to a spinal board on the rigid platform of a motion simulator.

The head is supported by a cushioned surface underneath and cushions on the sides (head pad) to support the head and reduce side-to-side motions.

3.3.2 Effect of vibration and immobilization on human biodynamic response

3.3.2.1 Effect of support surfaces

The following example demonstrates the effects of transport surface materials and conditions on the biodynamic response of supine humans. Fig. 3.23

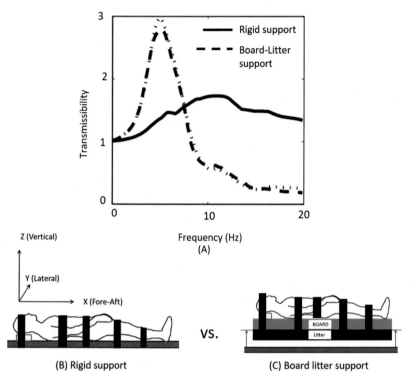

FIGURE 3.23 (A) Average of 12 subjects' responses at the sternum in terms of the transmissibility function in the vertical translational directions with humans on (B) rigid supports and (C) board litter supports. *Adapted from (A) Meusch, J. C., & Rahmatalla, S. (2014a). 3D transmissibility and relative transmissibility of immobilized supine humans during transportation.* Journal of Low Frequency Noise Vibration and Active Control, 33(2), 125—138. https://doi.org/ 10.1260/0263-0923.33.2.125.

provides a schematic representation of two transport support conditions, one in which the humans are laid on their backs on a rigid surface (Fig. 3.23B, which simulates laying the person on a rigid floor) and one in which the humans are laid on a spinal board and litter system (Fig. 3.23C) that is traditionally used during the evacuation and prehospital transport of injured patients with potential spinal cord injuries (Rahmatalla et al., 2020).

The graph in Fig. 3.23A shows the transmissibility graph in the vertical gravity direction (Z-axis) at the sternum of the person in response to the vertical input around the pelvic area at the interface in the vertical gravity direction. The biodynamic responses of supine humans under these two conditions show different characteristics. At low frequencies on the rigid surface, the graph shows a transmissibility with a magnitude of 1, indicating no magnification or reduction in the input motion; in this case, the output motion on the human body will be similar to the input motion to the body.

For the person on the rigid support, the graph starts to climb after 2—3 Hz and reaches a peak between 8 and 12 Hz, which indicates a resonance phenomenon. At this peak, the transmissibility value is almost 1.7, and therefore it is expected that the output motion at the chest will be almost 1.7 that of the input motion. The graph then descends. On the other hand, the graph for the person on the board—litter system shows a peak around 5 Hz and then descends sharply. After 8 Hz, the transmissibility value drops below 1, which means that the output motion is much less than the input motion and the person will hardly feel the input vibration. Although this may seem positive, the board—litter system can still generate a large peak at 5 Hz with a transmissibility value of around 3, meaning that the motion of the person will be three times greater than the motion entering their body. This should be a point of concern because most transport vehicles have suspension systems that can generate peaks around 2—5 Hz. This will generate very bad conditions for transported humans; they will suffer from double magnification (from the vehicle and from the transport system) if they are exposed to a road profile that generates frequencies of 3—5 Hz. It should be mentioned here that this example does not indicate that rigid surfaces are better than the board litter support, as rigid surfaces have other issues and are uncomfortable in general.

3.3.2.2 Effect of straps

Fig. 3.24 shows participants being tested during WBV using (B) a litter with no straps, (C) a litter with straps, (D) a litter and spinal board with no straps, and (E) a litter with spinal board and straps. Fig. 3.24A shows the magnitude of the acceleration at the head-to-chest (neck) and chest-to-hip (lower back) regions for the different immobilization configurations. Fig. 3.24A shows that the addition of straps in cases (C) and (E) does not significantly affect transmitted vibration from the corresponding no-strap conditions (B) and (D). However, the acceleration is significantly reduced in the vertical head-to-chest direction when the spinal board is added. While this indicates that the addition of the spinal board increases the mitigation capability of the litter system and the addition of straps does not significantly affect the outcome, other factors, such as the vibration magnitude involved and the effect of shocks, should be considered before jumping to a conclusion about the effectiveness of straps, as will be shown in the following sections.

3.3.2.3 Effect of shocks

Fig. 3.25A demonstrates the resulting vibration magnitude at the neck and lower back of a supine human exposed to shocks. Two immobilization conditions are considered in this example: case (B), where the person is immobilized with a litter with straps, and case (C), where the person is immobilized with a litter and spinal board with straps. As shown in Fig. 3.25A, the addition of straps with the spinal board and litter significantly reduced the

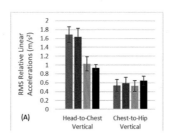

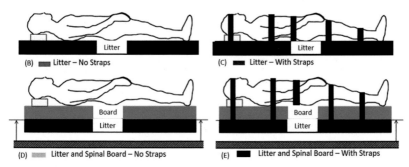

FIGURE 3.24   The resulting relative motions during WBV at the neck and lower back regions: (A) average acceleration values for participants under four cases with the error bars based on ± 1 standard deviation, (B) litter with no straps, (C) litter with straps, (D) litter with spinal board and no straps, and (E) litter with spinal board and straps. *Adapted from (A) Rahmatalla, S., DeShaw, J., & Barazanji, K. (2017). Biodynamics of supine humans and interaction with transport systems during vibration and shocks. Journal of Low Frequency Noise, Vibration and Active Control, 38(2), 1–9.*

transmitted vibration during shocks, and it is especially clear in the lower back region. In this case, the transmitted vibration at the lower back area between the two different immobilization conditions is reduced by more than 50% under shocks, where the transmitted vibration at the lower back area was almost similar when the shocks did not exist (Fig. 3.24A). While the addition of the spinal board has also improved the situation at the neck, as shown in Fig. 3.24A, the difference between the vibration in Fig. 3.24A and the shocks in Fig. 3.25A may show additional vibration mitigation to the neck region.

3.3.2.4   Effect of vibration magnitude

The literature shows that human biodynamic response to vibration can be affected by vibration magnitude and that human response will show nonlinear softening behavior (appearing less stiff and more relaxed) with increasing vibrational magnitudes (Huang & Griffin, 2008; Matsumoto & Griffin, 2002b; Rahmatalla & Liu, 2012). This is counterintuitive, as one might expect the body to become stiffer as vibration magnitude increases and

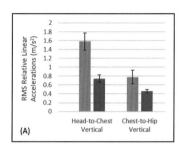

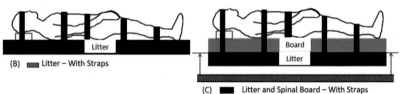

FIGURE 3.25 (A) The resulting relative motion at the neck and the lower back when using (B) litter with straps versus (C) spinal board with litter and straps during whole-body vibration with shocks. Bars indicate the average acceleration values for participants. The error bars are based on ± 1 standard deviation. *Adapted from (A) Rahmatalla, S., DeShaw, J., & Barazanji, K. (2017). Biodynamics of supine humans and interaction with transport systems during vibration and shocks.* Journal of Low Frequency Noise, Vibration and Active Control, 38(2), 1–9.

humans tense their muscles to resist the incoming vibration. However, researchers have found the nonlinear softening phenomenon in humans in seated, standing, and supine positions. These nonlinearity trends were seen at the head, torso, and pelvis. From the literature, one can summarize the source of nonlinearity (mostly softening behaviors) with increasing vibration magnitude as tissue thixotropy (the property of a material becoming less viscous when shaken, such as ketchup liquefying when the bottle is shaken) and voluntary and involuntary muscle activations at the local and global levels.

Fig. 3.26 shows the results of supine humans being exposed to vibration of increasing magnitudes (0.6, 1.0, and 1.3 m/s^2). Fig. 3.26 presents average transmissibility between the input and output motions at the (A) chest and (B) pelvis of 26 human subjects. Both graphs show the peaks in the transmissibility functions moving to the left at lower frequencies with increasing vibration magnitude, indicating a softening behavior. Fig. 3.26B shows that the magnitude of the peaks in the transmissibility functions are lower when the input vibration increases.

3.3.3 Relative transmissibility

Relative transmissibility is a useful function that can be used to investigate the biodynamic response at a certain region of the body (Meusch, 2012; Meusch & Rahmatalla, 2014a). The relative transmissibility between the

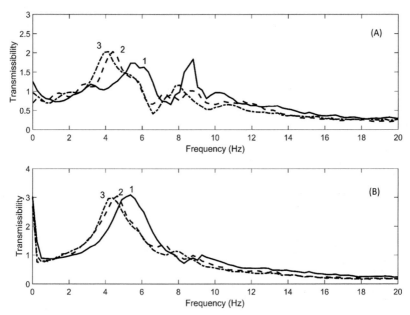

FIGURE 3.26 Average transmissibility between the vehicle floor and the (A) chest and (B) pelvis of 26 supine humans transported under three road profiles with root mean square vibration magnitudes of 0.5 m/s^2 (profile 1), 1.0 m/s^2 (profile 2), and 1.3 m/s^2 (profile 3).

head and chest is an example of what is happening at the neck level. Mathematically, the expressions for relative transmissibility $T_{rel}(f)$ are similar to those for traditional transmissibility. The difference here is that the output motion ratio between the two segments will be used instead of the input at the interface.

$$T_{rel}(f) = \frac{a_i(f)}{a_j(f)} \quad (3.12)$$

where $a_i(f)$ is the output acceleration at a body segment (i), such as the head, and $a_j(f)$ is the output acceleration at an adjacent segment on the human body, such as the torso.

In cases where the relative transmissibility is used to investigate the difference between systems with different input vibrations, the effect of the variations in the input motion can be normalized by introducing the following expression:

$$T_{rel}(f) = \frac{a_i(f) - a_j(f)}{a_{inp}(f)} \quad (3.13)$$

where $a_{inp}(f)$ is the input acceleration at the interface.

Fig. 3.27 shows an example of the relative transmissibility between the head and chest for subjects on (B) rigid and (C) board and litter support systems.

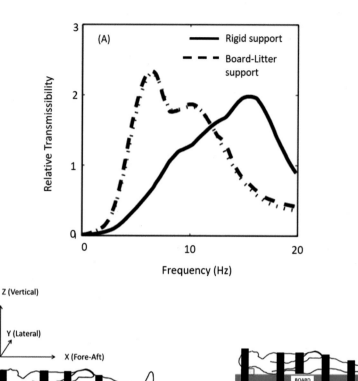

FIGURE 3.27 (A) Median platform-to-torso-head relative transmissibility of eight subjects on (B) a rigid support and (C) a board–litter support. *Adapted from (A) Meusch, J. C., & Rahmatalla, S. (2014a). 3D transmissibility and relative transmissibility of immobilized supine humans during transportation.* Journal of Low Frequency Noise Vibration and Active Control, *33(2), 125–138. https://doi.org/10.1260/0263-0923.33.2.125.*

Fig. 3.27A shows that the person on the rigid support can be exposed to double magnification of the input motion at frequencies around 17 Hz and that this is mitigated considerably when the person is transported on a board–litter system. The board–litter system, on the other hand, can magnify the relative transmissibility more than 2.5 times at 5 Hz and close to 2 times at 10 Hz. The biodynamic response of the supine human during transportation, especially the relative motion at the cervical region (between head and chest) and the lumbar region (between chest and pelvis), plays a significant role in a patient's safety and health outcomes. The relative motion at the cervical and lumbar regions can be affected by the direct input energy from the transport system to individual body segments (head, torso, pelvis, and legs), the energy transmitted from adjacent segments, and the motion coming from the whole body.

3.3.4   Case study: 3D transmissibility of supine subject

In this case study (Meusch, 2014a), eight healthy supine human participants were tested under SIP−SOP and 3IP−3OP random WBV. The 3IP vibration comprises white noise components with a total root mean square (RMS) magnitude of 1.0 m/s^2. The SIP vibration used 1.0 m/s^2 RMS in each translational direction. The vibration ride files were played using a six-degree-of-freedom man-rated shaker table (Moog-FCS 628). The subjects were tested under three different immobilization conditions: (i) subjects lying on a rigid platform and strapped to it, as shown in Fig. 3.28A; (ii) subjects lying on a spinal board strapped to a standard military litter, with the whole system

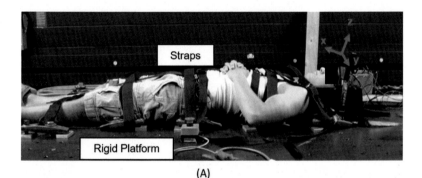

(A)

(B)

FIGURE 3.28 Supine human testing under single-input and multiple-input random whole-body vibration: (A) subject lying on a rigid platform, (B) subject lying on a spinal board placed on a litter, with the board−litter attached to the rigid platform. *Adapted from Meusch, J. C., & Rahmatalla, S. (2014a). 3D transmissibility and relative transmissibility of immobilized supine humans during transportation.* Journal of Low Frequency Noise Vibration and Active Control, *33(2), 125−138. https://doi.org/10.1260/0263-0923.33.2.125.*

attached to a rigid platform, as shown in Fig. 3.28B; and (iii) a condition similar to (ii) but with the addition of a neck collar. Inertial sensors were attached to the human body at the head, torso, pelvis, and thigh. Sensors were also attached to the rigid platform and the spinal board to measure the input motion at the interface between the human body and the contact surfaces. Inertial sensors attached to the back of the long spinal board were approximately below the head, torso, and pelvis. The inertial sensors on the human body were secured using foam wraps.

Fig. 3.29 shows the mean and standard deviation of the eight subjects at the head and torso at selected frequencies between 3 and 15 Hz. A comparison between the transmissibility components for the different cases showed that the transmissibility at the head and torso with the spinal board and litter

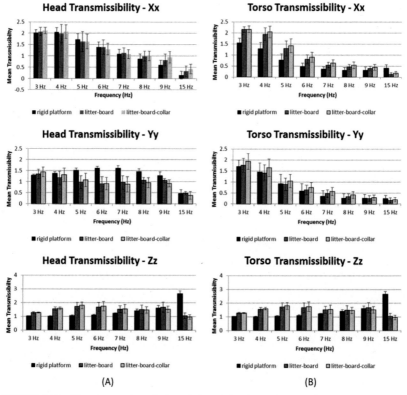

(A)   (B)

FIGURE 3.29 The mean and standard deviation of the diagonal components for the head and torso under different conditions (rigid platform, litter—board, and litter—board—collar) and selected frequencies. *Adapted from Meusch, J. C., & Rahmatalla, S. (2014a). 3D transmissibility and relative transmissibility of immobilized supine humans during transportation.* Journal of Low Frequency Noise Vibration and Active Control, *33(2), 125—138. https://doi.org/10.1260/ 0263-0923.33.2.125.*

for case (ii) overlapped with that of case (iii) but differed from the rigid case (i) at most frequencies, meaning that the neck collar did not have significant effects on the mean transmitted vibration. Meanwhile, the mean transmitted vibration under the rigid case (i) was lower than case (ii) and case (iii) under most frequencies, except 15 Hz at the torso. This could be related to the capability of the litter to mitigate the vibration at higher frequencies.

The results also showed comparable values between the transmissibility components of SIP−SOP in the X_x, Y_y, and Z_z directions and the corresponding diagonal components of the 3IP−3OP vibration (Fig. 3.30); the statistical differences between them showed a significant correlation ($R^2 > 0.93$). However, the correlations ($R^2 = 0.11−0.99$) were mostly weak between the cross-axis of SIP−SOP and the off-diagonal components of the 3IP−3OP conditions. The similarity between the individual direction of SIP−SOP and the corresponding diagonal components of the 3IP−3OP condition is demonstrated in Fig. 3.30, which shows the relationship at the sternum location on the torso. Fig. 3.30 also shows that there are differences between the individual subjects but that their means are very close in the three directions.

This case study presented that the transmissibility and relative transmissibility of a supine human under multiaxis input/output WBV showed a high correlation with that under single-axis input/output WBV, especially in the diagonal components. However, it showed a weaker correlation in the out-of-diagonal (cross-axis) components. This may suggest using single-axis testing to investigate multiaxis vibration, especially for the main diagonal components. The diagonal components of the multiaxis input/output transmissibility and relative transmissibility are considerably larger than their out-of-diagonal components.

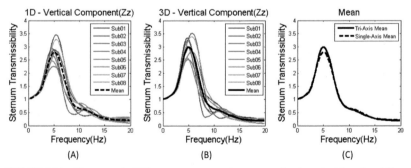

FIGURE 3.30 The transmissibility components at the sternum for the spinal board and litter condition: (A) the 1D vertical Z_z component for single input−single output, (B) the 3D vertical Z_z component for 3D-multiple input−3D-multiple output, and (C) the difference between the mean transmissibility in (A) and (B). *Adapted from Meusch, J. C., & Rahmatalla, S. (2014a). 3D transmissibility and relative transmissibility of immobilized supine humans during transportation.* Journal of Low Frequency Noise Vibration and Active Control, 33(2), 125−138. https://doi.org/10.1260/0263-0923.33.2.125.

3.3.5   Effect of posture

The literature on human response to vibration has demonstrated that the posture humans take can play a critical role in their biodynamic response (Kittusamy & Buchholz, 2004; Rehn et al., 2005). Nonneutral postures, such as awkward postures in WBV, can cause discomfort and potential injuries if they are maintained for a prolonged time (Eger et al., 2008; Rehn et al., 2005). While this has been heavily investigated for seated and standing humans, the effect of posture on people in supine positions has been overlooked. One reason behind this negligence could be that people think posture is irrelevant during supine transport as people simply lie on their backs in normal or neutral postures. However, this may not be the case, especially during the prehospital transport of injured patients, when supine humans could be immobilized in positions other than lying on their backs. With injuries or burns, for example, supine patients could be immobilized to lie on one side of their body or be in prone positions during transport. Healthy humans can also take nonneutral postures during transport while they are sleeping in train berths, for example. One main shortcoming of the current studies on the effect of posture on human response and comfort during WBV is that most have been conducted on healthy humans. We all know that discomfort and pain can be triggered in sick or injured patients as a result of small changes in their postures or the injured joint motions. Therefore more studies are needed to design better transport systems that take human postures into consideration for the safety and wellbeing of humans.

There are many occupations in which people need to use awkward or nonneutral postures so they can monitor equipment while both the human and the equipment are under vibration (Eger et al., 2008; Newell & Mansfield, 2008; Thuresson et al., 2005). Taking an abnormal posture can affect the body's response in different ways. For example, twisting the neck to one side will tighten the connection between the head and the upper torso and make the neck stiffer than when it is in a normal (neutral) position; this can increase the local natural frequencies at these areas and may increase the risk for injury. As a result of nonneutral postures, the relative motion between body segments at certain joints can sometimes reach their upper or lower passive limits, causing a considerable degree of discomfort in some instances. Researchers have realized the effect of posture on human response and risk evaluation in WBV (Hinz et al., 2002; Johanning et al., 2006; Kitazaki & Griffin, 1998). Considering seated posture, Nawayseh and Griffin (2005) investigated the dynamic responses at the seat and footrest of 18 subjects sitting on a seat with and without a backrest and with varying footrest heights. In this case study, the subjects took four different postures. The researchers found that the backrest reduced vertical and fore-and-aft forces at the footrest at frequencies below resonance. Also, they found that nonlinear behaviors of the forces on the seat and the footrest varied between postures.

In another study on seated subjects during vertical WBV, Hinz et al. (2002) conducted experiments on 39 male subjects sitting on suspension seats with and without backrests. They concluded that backrest and posture conditions did play an important role in the subjects' biodynamic response and therefore should be included in risk assessment during WBV. Wang et al. (2006) considered 36 different sitting postures and seat configurations and found a significant effect of sitting posture on the biodynamic response under vertical vibration. Their results showed important combined effects of inclined backrest and hand position on the AP characteristics. Although they are somewhat varied, most previous studies have demonstrated the importance of considering postures when investigating the human response to WBV.

As many critical injuries can happen to the head–neck area during accidents, natural disasters, and wars, investigating the effect of posture on the biodynamics of the neck should be seriously considered as it could play an important role during the transport of supine patients. In the general area of human response to vibration, the head-neck postures in WBV have received less attention when compared to the considerable focus on the lower back area of the spine. Rehn et al. (2005) conducted a review on the occupational use of all-terrain vehicles (ATVs). They showed that the nonneutral rotational positions of the neck are an ergonomic risk factor that occurs frequently and with a short duration for professional ATV drivers. The prevalence of serious neck and lower back disorders among locomotive engineers was found to be nearly double that of the sedentary control group without such exposure (Johanning et al., 2006). In another study, Courtney and Chan (1999) performed an ergonomic study to assess the workspace of grab unloaders for bulk materials in ships. They found that all drivers complained of neck and back discomfort while spending 50% of their time looking down vertically. In their investigation on 14 Swedish helicopter pilots who normally used neutral neck positions and neck flexing at 20 degrees, Thuresson et al. (2005) found that the neck position seemed to have a significant influence on the induced load and muscle activities. In their study on load-haul-dump drivers' risk of musculoskeletal injury in the mining industry, Eger et al. (2008) showed that when people were simultaneously subjected to vibration and nonneutral postures such as neck rotation and truck rotation, flexion, and lateral postures, it could increase their risk of injury.

3.3.6   Effects of gender, mass, and anthropometry

Studies in the area of human response to vibration have demonstrated that people are sensitive to the magnitude and frequency of the vibration. This sensitivity is also affected by the posture humans take, their muscle activity, and how they interact with the surrounding equipment. Research on humans exposed to WBV also showed that human response can be affected by gender, body mass, and anthropometry (Dewangan et al., 2013;

Griffin & Whitham, 1978; Nawayseh et al., 2019; Shibata et al., 2012; Toward & Griffin, 2011). In a study on seated people subjected to WBV, Dewangan et al. (2013) demonstrated clear gender differences. The study found that male subjects generated higher primary resonance frequencies than female subjects. Lundström et al. (1998) attributed these differences to the amount of fat in the body of females and suggested that fat as a damping material may absorb more energy than muscles. While they found clear differences in human response between genders, they still wondered if the differences were due to the mass of the subjects rather than their gender. However, when they normalized the response by the mass of the subjects, that is, removed the effect of mass from the response data, they found that there were still differences between the female and male subjects. In another important work on seated positions, Lundström et al. (1998) found a strong relationship between human response and body mass and gender. They also found that the largest peak in the response of female subjects took place at lower frequencies than in male subjects, which could indicate that male subjects demonstrated a stiffer body response. Additionally, they found that female subjects seemed to absorb more power than male subjects, which could be related to their body structure and the percentage of fat in their bodies; this is consistent with the finding of Dewangan et al. (2013). Lundström et al. (1998) also reached a similar conclusion and suggested a need for differentiated guidelines for female and male subjects during transport; however, they were hindered by the limited number of subjects they tested.

Very little work has been done to investigate the effects of gender, mass, and anthropometry in supine positions under WBV (Rahmatalla et al., 2020). In addition, most studies on human response to WBV have focused on the translational motions (up-down, side-to-side, and fore-aft) in human response; few have paid attention to how the segments of the body rotate (roll, pitch, and yaw) when the person is subjected to WBV. It should be mentioned here that while some works have investigated the effect of rotational input motion on the biodynamics of seated and standing subjects in WBV (Paddan & Griffin, 1994; Parsons & Griffin, 1978), very few have been done to investigate the effect of rotational vibrations on the biodynamic response of humans due to gender, mass, and anthropometry (Rahmatalla et al., 2020). Investigation into the motion of human segments under WBV in the six translational and rotational directions (all types of motions humans can generate at their joints) will provide valuable information for understanding how humans respond to vibration. For example, studies by DeShaw & Rahmatalla (2014b, 2016a, 2016b) have shown that human discomfort is more related to the rotational velocities of the segments than to the translational accelerations. The information gained from such studies will provide more understanding to human response to vibration and may ensure human safety and lead to improved prehospital transport system designs.

Although this book focuses on supine humans in prehospital transport, the effects of gender, body mass, and anthropometry on human response to WBV have mainly been studied in seated positions. Therefore some details from these works will be used for comparison purposes. There are two field studies that are recently published (Rahmatalla et al., 2020, 2021) in which the effects of gender, body mass, anthropometry, and vibration magnitude on supine-human response to WBV were investigated. The results from these two articles provide the primary basis for discussion in the remainder of this chapter.

The following sections investigate the effect of gender, body mass, and stature of a human in the supine position under WBV with special emphasis on the effect of body-segment geometry on the resulting rotational motions of these segments. They also demonstrate the effect of increasing vibration magnitudes on the behavior of the human body and how the human body becomes softer and stiffer under different vibration magnitudes. Field data under road conditions, similar to a certain degree to those taking place in real-life scenarios with extremely rough conditions such as medical evacuations in theater, were used in this work, and acceleration-based and angular velocity-based transmissibility analyses were performed. The following sections discuss the preparation, data collection, and data interpretation during a simulated prehospital transport field study.

3.3.7   Case study: example field study

This section demonstrates traditional procedures used in the military to immobilize and transport patients in ground ambulances. The immobilization procedures are generic and are applied to everyone regardless of their gender, body mass, and anthropometry. This example will show the differences between the biodynamic responses of different subjects when immobilized with the same system and investigate whether or not the differences are statistically significant.

3.3.7.1   Subject preparation

In this example (Rahmatalla et al., 2020, 2021), 26 healthy subjects with no musculoskeletal conditions participated in a simulated field test. The subject population included 14 males and 12 females with an average body mass of 81.9 kg (standard deviation of 15.6 kg) and an average body height of 1.731 m (standard deviation of 0.082 m). The male subjects had an average body mass of 90.2 kg (standard deviation of 13.6 kg) and an average body height of 1.777 m (standard deviation of 0.066 m). The female subjects had an average body mass of 72.2 kg (standard deviation of 12.0 kg) and an average body height of 1.678 m (standard deviation of 0.064 m). The words *heavy* and *light* are used in this example to refer to subjects who had higher than average body mass and lower than average body mass, respectively. The words *tall* and *short* are used to refer to subjects who had higher than average body height and lower-than-average body height, respectively.

The subject preparation and immobilization techniques used in this example represent standard procedures used at the USAARL. During the preparation stage, the human subject was laid on a spine board. A head cushion was placed under the subject's head, and support blocks were attached to the cushion to prevent side-to-side movements of the head. Standard straps were used to constrain the subject to the spine board. After this preparation stage, sensors were attached to the subject. In this example, six-degree-of-freedom inertial sensors (Dytran Instruments, Inc., Chatsworth, CA, United States) were used instead of traditional accelerometers because, as mentioned in Chapter 2, inertial sensors can measure three-directional linear translational accelerations and three-directional rotational angular velocities. Six sensors were used: one was attached to the floor of the ambulance vehicle and one to the spine board, approximately under the pelvis of the subject, to measure the input motion to the subject body, and the rest were attached to the subject's head, chest, pelvis, and knee to measure body motion. Double-sided tape was used to attach the sensors to the human body. The sensors were also secured to the subject's body using banded strips of athletic and medical tape to reduce skin-on-bone motions. A data acquisition system (Crystal Instruments CoCo-90 Dynamic Signal Analyzers) was used to record motion at 500 samples per second. Then the spine board with the human on it was placed on a standard United States Army litter. Another set of straps was used to secure the human and spine board to the litter. Finally, the litter was placed inside an M997 high-mobility multipurpose wheeled vehicle ground ambulance and attached firmly to its frame.

3.3.7.2   Testing and data collection

During testing, the ambulance was driven at different speeds (16−24 km/h) on a road that resembled real-life off-road transport conditions. The track is a loop (Fig. 3.31) and has seven sections: (Section 1) gravel; (Section 2) outside track (bumpy), (Section 3) outside smooth; (Section 4) inside track (bumpy), (Section 5) inside smooth; (Section 6) straddle; and (Section 7) half-rounds, where half-round steel pipes were fixed to the ground, simulating a bump in the road and generating shock-like conditions. The collected data were postprocessed using a low-pass filter at 40 Hz.

Fig. 3.32 shows an example, using Subject 15, of the vertical vibration profiles at the vehicle floor during one turn around the track.

Fig. 3.32B shows that the acceleration signals in the time domain and the magnitude of acceleration changed from section to section of the driving terrain, with more severe vibrations in Sections 2, 4, and 6. Another way to determine the severity of a vibration signal is to look at the RMS value, which represents one number, conceptually similar to the average value, that reflects the severity of vibration. For example, the RMS value for Section 2 is 5.5755 m/s^2 and for Section 3 is 0.982 m/s^2. Another option is to look at the vibration signals'

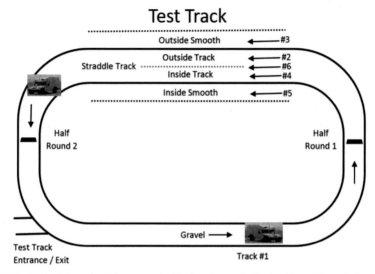

FIGURE 3.31 Schematic of the test track, (Section 1) gravel; (Section 2) outside track (bumpy), (Section 3) outside smooth; (Section 4) inside track (bumpy), (Section 5) inside smooth; (Section 6) straddle, two tiers on inside bumpy and two tires on outside bumpy; and (Section 7) half-rounds, where half-round steel pipes were fixed to the ground.

power spectral density (PSD), which shows their power at each frequency. In Fig. 3.32C, the PSD graphs of the different sections show that the majority of the energy in the low-vibration signals (Sections 1, 3, and 5) is around 2 Hz. It then slowly descends before showing another prominent peak around 10 Hz. The energy in sections with higher vibration magnitudes (Sections 2, 4, and 6) is mainly concentrated around 2−3 Hz and then quickly drops.

3.3.8 Data analysis

In this example, the data is analyzed using two approaches. The first approach is to use the transmissibility function, which quantifies the magnification of the magnitude between the input vibration to the vehicle and the output vibration at segments of the human body such as the head, chest, and pelvis. The resulting transmissibility function normally shows the relationship and magnification between the input and output motions at each frequency. From the transmissibility graphs, it would be possible to see the frequencies at which the largest magnification is taking place. Traditionally, during lab testing the transmissibility graphs are smooth and the frequencies where the transmissibility shows peaks are clear. In field studies, however, because of the noise and low energy at some input frequencies, the transmissibility function will not be smooth across all frequencies and instead will have many artificial peaks. With these artificial peaks, comparing the

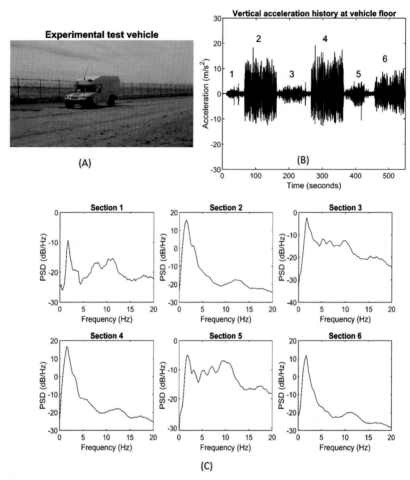

FIGURE 3.32 (A) Experimental test vehicle, (B) vertical acceleration history at vehicle floor, and (C) average power spectral density of vertical acceleration at vehicle floor for Sections 1–6.

different cases becomes challenging. To circumvent the effect of such numerical anomalies, another approach was used in the analysis. Rather than the peak frequencies, the area under the transmissibility graphs was used to compare the cases. It should be mentioned that analysis based on the peak frequency will be also used when the peaks are prominent in the graphs.

In this work, the investigation of the effects of gender, mass, stature, and vibration magnitude on the biodynamic response of a supine human is presented in terms of the major peak frequency in the transfer function if the peaks are obvious. These transfer functions will be constructed using translational acceleration (Dewangan et al., 2013; Lundström et al., 1998; Toward & Griffin, 2011) in the vertical gravity direction and rotational velocities in the roll and pitch directions. Otherwise, it will be investigated in terms of the area under the transmissibility graphs. The transmissibility in general represents the energy through the body as it relates to the input and output motion, as the expectation is that the energy will flow inside the body between the input and output locations. The area under the transmissibility graphs gives a good representation of the energy going through the body, and it may provide useful information about the difference in the response between different subjects. It should be noted that the transmissibility graphs resulting from the angular velocities were noisier than those from the acceleration; therefore only the areas under the transmissibility graphs of the angular velocities will be used in the analysis. The peaks in the transmissibility graphs will be used only when they are very distinct; otherwise, the area under the transmissibility graph, which represents the region between the line representing the transmissibility graph and the horizontal axis in the graph, will be used. The angular velocities will be discussed in terms of the rolling and pitching motions of the body segments. For the head, a rolling motion can be defined as a side-to-side motion, as when a person says "no." A pitching motion is similar to the nodding gesture of the head when a person says "yes."

3.3.9   General findings

This section discusses the findings regarding the effects of gender, body mass, stature, and vibration magnitude on the locations of the peaks in the transmissibility graphs and the areas under the transmissibility graphs for different subjects. One interesting issue that will be discussed in this section is the effects of angular motion and the shape and dimension of the human body segments on the biodynamical response during vibration. The angular motions have been shown to play a major role in human discomfort in WBV (DeShaw & Rahmatalla, 2016a, 2016b), but unfortunately, they have been overlooked by researchers. Additionally, it is well known that with increasing vibration magnitude, the human response becomes nonlinear, meaning that the body becomes softer and less stiff. Debates on the nonlinear response of the human body to vibration are ongoing and have centered on the amount of fat in the body, the thixotropy of human tissues, and voluntary and involuntary motions, among other factors. The field study presented here revealed a very interesting phenomenon in which supine humans showed softening behaviors at increasing low-vibration magnitudes but stiffening behaviors when the vibration magnitudes become higher.

The results of this work also show that gender, mass, stature, and vibration magnitude can have a significant effect on how supine humans respond to WBV. This is consistent with what has been seen in seated and standing subjects (Dewangan et al., 2013; Griffin & Whitham, 1978; Nawayseh et al., 2019). Because of the severity of the road conditions and vibration in this work, the human subjects experienced considerable translational and rotational motions. The rotational motions showed strong relationships with the geometry of the body segments and varied in magnitude depending on gender, stature, and body mass. This is an important finding and should be considered by transport system developers as most immobilization systems are designed to reduce motions in the translational directions, mostly in the vertical gravity direction, and not in the rotational motions.

3.3.10   Effect of gender

Various segments of the human body can be differently affected by vibration depending on their natural frequencies and their distances from the input vibration source. The head−neck and lower back−pelvis regions are considered in most studies as the critical areas during transport of patients with traumatic and spinal cord injuries; therefore the focus of this section will mainly be on these two areas. The resulting vertical acceleration−based transmissibility graphs of the females and males at the pelvis depicted two prominent peaks at frequencies around 5−10 Hz, meaning that the pelvis could generate high motion if exposed to vibration containing these frequencies. Similar characteristics have been seen in lab experimentation and mathematical modeling of seated positions (Boileau & Rakheja, 1998; Holmlund & Lundström, 2001; Liang & Chiang, 2008; Qiao, 2017; Wang & Rahmatalla, 2013; Wei & Griffin, 1998). As shown in Fig. 3.33, the peak frequency of the average of the female subjects has a larger magnitude than that of the male subjects. The main dominant peak for females also occurred at lower frequencies, that is, was shifted to the left. Again, this is consistent

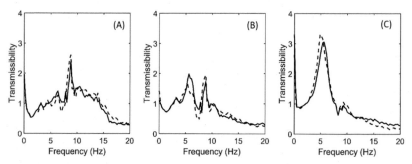

FIGURE 3.33   The transmissibility of vertical acceleration of the (A) head, (B) torso, and (C) pelvis relative to the vehicle floor in Section 1; solid lines represent males, and dashed lines represent females. *Adapted from Rahmatalla, S., Kinsler, R., Qiao, G., DeShaw, J., & Mayer, A. (2020). Effect of gender, stature, and body mass on immobilized supine-human response during en route care transport.* Journal of Low Frequency Noise, Vibration and Active Control. *https:// doi.org/10.1177/1461348420911253.*

with the findings (Dewangan et al., 2013; Lundström et al., 1998) in seated subjects when the AP and the AM were used. It should be mentioned here that the transmissibility graphs at the pelvis of both female and male supine subjects demonstrate a large peak at 5 Hz and a smaller peak around 10 Hz, which is similar to what has been observed using AP with seated humans.

It is interesting to note that human models for seated positions (Liang & Chiang, 2008; Qiao, 2017; Wang & Rahmatalla, 2013; Wei & Griffin, 1998) have also predicted the existence of two peaks at locations similar to those in Fig. 3.33. When looking at the area under the vertical acceleration-based transmissibility of the pelvis, the female subjects demonstrated higher magnitudes than male subjects. This means that the female subjects generated higher motions than male subjects in the vertical direction. This could also be attributed to the effect of mass as subjects with heavier weight (males) would generate less vertical motion. The effect of mass on the resulting motion has been discussed by other researchers (Toward & Griffin, 2011), and the results are consistent with those of this work.

The heads of the female subjects showed lower vertical translational vibration than those of the male subjects, especially on the bumpy segments of the road. This could be because females' heads are lighter than males' heads (Churchill et al., 1978; Department of Defense, 2012). Fig. 3.34 shows the resulting transmissibility graphs of the head, torso, and pelvis based on the angular motions in the roll and pitch directions. It is evident that the angular velocity−based transmissibility graphs are noisier than those for the translational accelerations and that there are many artificial peaks, which may result from the noisy data or from the transformation from the time domain to the frequency domain. Therefore the area under the transmissibility graphs will be used to compare the different cases.

The rotational motions of the females' heads showed overall relatively larger pitch motions (Fig. 3.34D) than those of the male subjects. One reason is that anthropometric studies (Churchill et al., 1978; Department of Defense, 2012) have shown that females have shorter faces/heads than males. Although female subjects have narrower heads, the results of this work demonstrated lower roll motions for females than for males in the low frequency range (Fig. 3.34A). While the expectation was that females with narrower heads would generate larger rolling head motion, the cushioned block that supports the head on both sides may have played a role in restricting the rolling motion of the head.

The human response in terms of angular velocities has provided very useful information. For example, the pelvis of female subjects showed higher angular velocities in the pitch direction (Fig. 3.34F) than that of the male subjects; however, it showed less angular velocity in the roll direction (Fig. 3.34C). While mass could be part of the reason, it is possible that it is due to differences in the anatomy and the geometry of female and male pelvises. Based on previous studies (Churchill et al., 1978; Department of

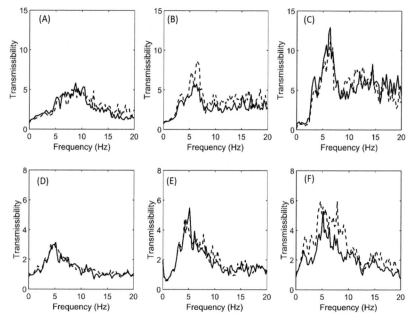

FIGURE 3.34   The transmissibility of roll and pitch angular velocities relative to the vehicle floor in Section 2: (A) head roll, (B) torso roll, (C) pelvis roll, (D) head pitch, (E) torso pitch, and (F) pelvis pitch; solid lines represent males and dashed lines represent females. *Adapted from Rahmatalla, S., Kinsler, R., Qiao, G., DeShaw, J., & Mayer, A. (2020). Effect of gender, stature, and body mass on immobilized supine-human response during en route care transport.* Journal of Low Frequency Noise, Vibration and Active Control. *https://doi.org/10.1177/1461348420911253.*

Defense, 2012; Wang et al., 2004), females in general have wider and shorter pelvises than males. As shown in Fig. 3.34C, females had a lower peak, at 6 Hz, at the pelvis in the roll direction. Meanwhile, females showed higher peaks than males at pelvis pitch (Fig. 3.34F) between 6 and 10 Hz. Females also showed more torso roll (Fig. 3.34B); this was statistically significant for Sections 4 and 6, where relatively severe inputs of rotational vibration took place. Coming back to geometry, the higher torso rolling in females could be due to females having narrower shoulders than males (Churchill et al., 1978; Department of Defense, 2012).

3.3.10.1 Effect of body mass

It was found that human subjects with high body mass (heavy) demonstrated lower vertical motion than those with low body mass (light). The behavior seen can be explained as the higher mass acting as a mass damper (Toward & Griffin, 2011). These results are also consistent with what Dewangan et al. (2013) found in their investigation on the effect of mass on seated humans. Videos and naked-eye observations during the field testing of supine humans showed that lighter subjects bounced and slid around the transport system more intensely than heavier subjects. The first major peak in the pelvis transmissibility was around 6 Hz for the light subjects, which is greater than that of the heavy subjects. Also, the frequency where the peak took place for the heavy subjects was lower than that of the light subjects; the peak of the heavy subjects was shifted slightly to the left. While the differences between the heavy and light subjects were clear for the pelvis, the nonsmooth behavior of the angular velocity-based transmissibility makes it harder to determine the difference between the segments.

The heavy subjects also showed more pelvis rolling and pitching motions than the light subjects. This was clear when the vehicle was subjected to extensive rolling and pitching motions when it was driven on some segments of the bumpy field terrain. In such circumstances, the litter is expected to be less effective against rotational input motions, as it is normally designed to mitigate vibration in the translational vertical direction. The greater rotational motions in heavy subjects could also be related to the type of straps used in the immobilization systems. In this case, one strap was used to hold the pelvis area, and it may not be effective in reducing the resulting rotational motion of the pelvis due to severe motions. The heads of the heavy subjects, however, showed lower translational motion than those of the light subjects, possibly because heavier subjects may have heavier heads than light subjects. In the immobilization system used in this field study, the head was supported by side blocks to prevent rolling motions; therefore the difference between the head rolling and pitching motions between the heavy and light subjects was not significant.

3.3.10.2 Effect of stature

Because of the noisy data in field testing, the transmissibility graphs were not smooth but rather showed zigzagged lines along frequencies, making it very hard to distinguish between the cases. Therefore the area under the transmissibility graphs is expected to smooth out these irregularities and present a reasonable comparison. The area under the vertical acceleration-based transmissibility at the torso and pelvis for tall and short subjects showed comparable results. This could be explained by the fact that some of the tall subjects were relatively heavy, and their mass could have played a role in their responses. Another reason is that the data used for stature comparison were taken from female and male subjects, and gender may have played a role as well. However, the angular velocity-based transmissibility showed that tall subjects had less torso roll than short subjects. This could be explained by investigating subjects' anatomy; tall subjects may have wider shoulders than short subjects (Churchill et al., 1978; Department of Defense, 2012). One thing to note is that the rolling motion of subjects with wider shoulders could be affected because their shoulders might touch the litter poles during transport.

Tall subjects, in general, showed less pelvic pitch than short subjects. This could be explained by their anatomy because tall subjects are expected to have longer pelvises than short subjects (Churchill et al., 1978; Department of Defense, 2012). It could also be because most of the tall subjects in this work were males with long pelvises. Tall subjects also showed less pelvis rolling motion than short subjects, but these results were not consistent across all the terrain sections used in this study. Tall subjects, in general, showed more rolling motion at the head than short subjects, but it was hard to find differences between the pitching motions, which could be attributed to the way the head was immobilized with side head supports.

3.3.10.3   Effect of vibration magnitude

It has been demonstrated that different road sections generate input vibrations of different magnitudes to the vehicle, and this is expected to introduce a softening behavior with increasing vibration magnitude to the biodynamical response of subjects due to nonlinearity and softening behaviors (Hinz et al., 2006; Huang & Griffin, 2009; Mansfield & Griffin, 2000; Rahmatalla & Liu, 2012). This means that subjects will show lower peaks with higher vibration magnitudes. Also, the peaks will shift to lower frequencies, indicating a system with lower stiffness. The results of this work demonstrated that the response of the subjects varied between softening during the sections with relatively low vibration magnitudes and stiffening during the sections where vibration magnitudes were relatively higher (Rahmatalla et al., 2021). This is a very interesting phenomenon that has not been seen in previous publications. The magnitude of the input vibration and the intersubject variability have also created some uncertainty in deciding if there is a significant difference between some cases. This can happen when the magnitudes of the subjects' averages have comparable values and when the standard deviations are relatively high. A situation like this can be seen in the torso roll.

As expected, and in line with the literature, the transmissibility graphs of supine human subjects showed nonlinear softening behaviors when they were exposed to WBV of increasing magnitudes and when the input vibrations were lower than 3 m/s^2. However, their responses showed stiffening behaviors with increasing vibration magnitudes when the vibration magnitudes exceeded 3−5 m/s^2. These nonlinearity trends were seen at the head, torso, and pelvis of the human subjects. While softening behavior is likely and is similar to what has been reported in the literature for seated, standing, and supine positions using different types of transfer functions (Dewangan et al., 2018; Fairley & Griffin, 1989; Hinz et al., 2006; Huang & Griffin, 2009; Lundström et al., 1998; Matsumoto & Griffin, 2002a, 2002b; Qiu & Griffin, 2004; Smith & Kazarian, 1994), nonlinearity with dominant stiffening behaviors has not been reported for humans under increasing vibration magnitudes. It should be mentioned, however, that stiffening behaviors have been reported in humans subjected to increasing constant acceleration magnitudes (Vogt et al., 1968), but not with increasing vibration magnitudes.

Many articles, reports, and theses have investigated the source of nonlinear behavior in the human response under increasing vibration magnitude (Huang, 2008; Matsumoto & Griffin, 2002a; Rahmatalla & Liu, 2012). Most researchers have agreed that the source of nonlinear softening behavior has been related mostly to tissue thixotropy (Huang & Griffin, 2009; Mansfield et al., 2006). Some researchers, on the other hand, related the nonlinear behavior to the effect of posture and the voluntary and involuntary muscle properties and activation. In one article, Matsumoto and Griffin (2002a) investigated the effect of muscle tension on the resulting nonlinearity in the AM and found relatively less softening when the seated subjects tensed their buttocks and abdominal muscles. Similar findings were demonstrated by Mansfield et al. (2006); they noticed a decrease in the softening of the AM when the subject used a tense posture rather than a relaxed posture. The findings from the latter articles may suggest that local and some major muscle activations may generate stiffening behaviors that may reduce the softening effect of the tissue thixotropy at these regions. Large muscle tonic activation, on the other hand, such as that with static acceleration (Vogt et al., 1968), may change the response characteristic of the body by creating a stronger connectivity between its segments and producing a global stiffening rather than a softening behavior.

The source of nonlinearity (mostly softening behaviors) with increasing vibration magnitude in the literature can be summarized as tissue thixotropy and voluntary and involuntary muscle activations at the local and global levels. Most current findings (Rahmatalla et al., 2021), however, demonstrate that stiffening behaviors can also take place under higher vibration magnitudes that often occur in the field during en route care. In fact, videos taken inside an Army ambulance showed that human subjects braced their whole bodies when the road became bumpier under higher vibration magnitudes. This stiffening behavior could be similar to what happens when humans are subjected to increasing static acceleration, as demonstrated by Vogt et al. (1973). The results showed the disappearance of the second peak in the transmissibility graphs at the torso and pelvis when the subjects were subjected to higher vibration magnitudes. The number of peaks in the transmissibility graph reflects the number of masses in the system. So if there are two peaks in the transmissibility graph, that may indicate a system of two degrees of freedom with two masses. When the second peak disappears, that may indicate that the system becomes a single degree of freedom with one mass, which may be related to the stiffening effect when major muscle activations hold the masses of the body together to resist the motion coming from vibration.

The transmissibility graphs of the female and male subjects showed similar behaviors under increasing vibration magnitudes: softening characteristics at RMS vibration magnitudes below 3 m/s^2 and stiffening behaviors at RMS vibration magnitudes of more than 3 m/s^2. In spite of the uncertainty in the data collections, due to the high vibration magnitudes and the severe motions in the six-directional space, the transmissibility graphs showed repeatable behaviors for most cases. Based on these results, it can be recommended that the vibration magnitude be considered in the development of transport systems for various environments. The results presented here can also have significant implications on supine patients who are conscious or those who are not compliant under the immobilized conditions and may generate additional unwanted motions and tonic muscle activations during transport (McDonald et al., 2020). This could also be true for a patient who is conscious and in pain, thus exhibiting more rigid muscle tone. These findings may also affect the wellbeing of unconscious subjects who are less likely to exhibit tonic muscle response, but more work is necessary to see if these results are reproducible.

3.3.10.4 Limitations

The results of these field studies (Rahmatalla et al., 2020, 2021) provide a lot of information for the field of supine transport of humans and therefore provide many benefits to the field of prehospital and en route transport. However, there were some limitations and assumptions in the analysis of the translational data that should be considered in potential applications. The situation when using the angular velocities, however, is different than the translational accelerations. In this case, as the angular velocity at the interface between the subject and the transport system is very similar and the subject and the immobilization system can rotate together, it may make more sense to use the angular velocity at the litter berth to which the patient is attached. Another limitation of these results is that subjects with wider shoulders may touch the litter poles with their shoulders, which could restrain their movements in the roll direction. In an extreme case, the interference between the shoulders could prevent rolling motion altogether. In addition, because the subjects were attached to a relatively rigid spine board with molded runners, heavy subjects may have pressed the runners deeper into the deformable mesh of the litter's fabric, which may have arrested the rolling motions.

3.4 Summary

This chapter summarizes the basic principles of the biodynamics of supine humans when subjected to WBV, as well as factors that affect biodynamics. It introduces the transfer functions used to quantify the biodynamic response and presents them in simplified mathematical formats. Because the transmissibility function is easy to measure and can provide valuable insights, it is discussed in more detail. With multiple-input and multiple-output vibration, the calculation and interpretation of the transmissibility function can become more involved; the transmissibility function can have up to 36 graphs in the most general case. The introduction of the effective transmissibility function is one way to simplify the transmissibility function for multiple-axis vibration and construct graphs similar to those in the SIP−SOP transmissibility.

Applications that affect the biodynamics of supine humans during transport, including the type of padding materials and the use of straps, are investigated with practical examples. The effects of increasing vibration magnitudes on the biodynamics of supine humans are also discussed using examples from a field study. The use of relative transmissibility to measure the relative motion between adjacent segments can be very helpful when investigating the efficacy of immobilization systems at certain regions of the human body. For example, it can be used to evaluate the effectiveness of the head support systems and cervical collars traditionally used when transporting patients with suspected spinal cord injuries.

The examples in this chapter show that the addition of a spinal board to the litter, for example, reduces the relative involuntary motion at the neck and torso regions. The results also show that the addition of a spinal board with body straps has either a neutral or a positive effect on the resulting human motion. As shocks with higher magnitudes are introduced, the benefit of adding a spinal board and straps becomes obvious. Intuitively, one would expect that adding more body straps would reduce relative motion between adjacent segments of the supine human body; however, the results show that this may not occur as expected. The relative linear accelerations do not show consistency, increasing in some directions and decreasing in others. In general, the combined effect of body straps and a spinal board, under the conditions considered in this chapter, effectively reduce the resulting motion on the human body, especially when shocks are present. The results presented here may be used for a general understanding of the effects of immobilization and transport systems on the biodynamic response of supine humans, but more work should be done to investigate systems in the field conditions in which they are used. While the research outlined in this chapter highlights the importance of evaluating the performance of transport and immobilization systems in environments that simulate real-life conditions, the performance of most transport systems on the market has not been evaluated in a dynamic setting where WBV exists; most of the work has been performed using static or very slow quasistatic conditions.

In this chapter, the effects of gender, body mass, and stature on supine-human response in a simulated field transport environment are also demonstrated. The chapter shows the effect of increasing vibration magnitudes on the biodynamic response of supine humans. While the examples in this chapter show similarity between the findings of seated and standing positions to those of supine positions, they also provide additional important information about the role of rotational motions and the effect of body shape when WBV exists. The relationship between segment geometry and the resulting rotational motions is rarely discussed in the context of supine-human response to vibration. The examples in this chapter show that, in general, a person with a large mass will experience reduced vertical motion of body segments during WBV but may experience increased torso and pelvis rotations. Female subjects can generate more pelvis pitching and translational motions, which could be attributed to the shape of their body segments and mass. This type of biodynamic response may play an important role in the safety and wellbeing of female patients with pelvic fracture. The examples also show that a person's stature may affect how much the torso rolls; however, they do not illuminate significant differences, suggesting the need for more laboratory and field studies. The results in general suggest that more care should be taken when immobilizing patients of different genders, body mass, and stature. One suggestion is to add lateral supports to the chest to reduce its rotation or cross-body supports to the pelves of people with spinal or pelvic injuries to mitigate unwanted translational and rotational motions. In terms of vibration magnitudes, the data presented in this chapter have shown obvious changes in the human response to vibration with increasing vibration magnitudes and the transformation from softening to stiffening behaviors with higher vibration magnitudes. While additional research is required, the results and discussions presented in this chapter may encourage those working in the area of prehospital transport to consider different immobilization systems and techniques when transporting people with different genders, body mass, and stature.

References

Boileau, P. É., & Rakheja, S. (1998). Whole-body vertical biodynamic response characteristics of the seated vehicle driver: Measurement and model development. *International Journal of Industrial Ergonomics, 22*(6), 449−472.

Churchill, E., Laubach, L. L., Mcconville, J. T., & Tebbetts, I. (1978). *Anthropometric source book: Anthropometry for designers (NASA Reference Publication 1024)* (Vol. I). Yellow Springs, OH: Webb Associates.

Coermann, R. R. (1962). The mechanical impedance of the human body in sitting and standing position at low frequencies. *Human Factors, 4*(5).

Courtney, A. J., & Chan, A. H. S. (1999). Ergonomics of grab unloaders for bulk materials handling. *International Journal of Industrial Ergonomics, 23*, 61−66.

Demic, M., & Lukic, J. (2009). Investigation of the transmission of fore and aft vibration through the human body. *Applied Ergonomics, 40*(4), 622−629.

Department of Defense. (2012). *MIL-STD-1472G, Department of Defense design criteria standard: Human engineering.* Arlington, VA: United States Department of Defense. (11.01.2012).

DeShaw, J., & Rahmatalla, S. (2014a). Effective seat-to-head transmissibility under combined-axis vibration and multiple postures. *International Journal of Vehicle Performance, 1*(3/4), 235. Available from https://doi.org/10.1504/IJVP.2014.069108.

DeShaw, J., & Rahmatalla, S. (2014b). Predictive discomfort in single- and combined-axis whole-body vibration considering different seated postures. *Human Factors, 56*(5), 850−863.

DeShaw, J., & Rahmatalla, S. (2016a). Predictive discomfort of supine humans in whole-body vibration and shock environments. *Ergonomics*, *59*(4), 568–581.

DeShaw, J., & Rahmatalla, S. (2016b). Effect of lumbar support on human-head movement and discomfort in whole-body vibration. *Occupational Ergonomics*, *13*(1), 3–14.

Dewangan, K., Rakheja, S., & Marcotte, P. (2018). Gender and anthropometric effects on whole-body vibration power absorption of the seated body. *Journal of Low Frequency Noise, Vibration and Active Control*, *37*, 167–190.

Dewangan, K. N., Shahmir, A., Rakheja, S., & Marcotte, P. (2013). Seated body apparent mass response to vertical whole-body vibration: Gender and anthropometric effects. *International Journal of Industrial Ergonomics*, *43*(4), 375–391.

Eger, T., Stevenson, J., Callaghan, J. P., Grenier, S., & VibRG. (2008). Predictions of health risks associated with the operation of load-haul-dump mining vehicles: Part 2—Evaluation of operator driving postures and associated postural loading. *International Journal of Industrial Ergonomics*, *38*, 801–815.

Fairley, T. E., & Griffin, M. J. (1989). The apparent mass of the seated human body: Vertical vibration. *Journal of Biomechanics*, *22*, 81–94.

Griffin, M., & Whitham, E. (1978). Individual variability and its effect on subjective and biodynamic response to whole-body vibration. *Journal of Sound and Vibration*, *58*(2), 239–250.

Griffin, M. J. (1990). *Handbook of human vibration*. London: Academic Press.

Heath, M. T. (1997). *Scientific computing: An introductory survey*. Boston: McGraw-Hill Inc.

Hibbeler, R. C. (2008). *Mechanics of materials* (7th ed.). New Jersey: Pearson Prentice Hall.

Hinz, B., Blüthner, R., Menzel, G., Rützel, S., Seidel, H., & Wölfel, H. P. (2006). Apparent mass of seated men—Determination with single- and multi-axis excitations at different magnitudes. *Journal of Sound and Vibration*, *298*(3), 788–809.

Hinz, B., & Seidel, H. (1987). The nonlinearity of the human body's dynamic response during sinusoidal whole-body vibration. *Industrial Health*, *25*, 169–181.

Hinz, B., Seidel, H., Menzel, G., & Bluthner, R. (2002). Effects related to random whole-body vibration and posture on suspended seat with and without backrest. *Journal of Sound and Vibration*, *253*, 265–282.

Holmlund, P., & Lundström, R. (2001). Mechanical impedance of the sitting human body in single-axis compared to multi-axis whole-body vibration exposure. *Clinical Biomechanics*, *16*(1), S101–S110.

Holmlund, P., Lundström, R., & Lindberg, L. (2000). Mechanical impedance of the human body in vertical direction. *Applied Ergonomics*, *31*(4), 415–422.

Huang, Y. (2008). *Mechanism of nonlinear biodynamic response of the human body exposed to whole-body vibration* (Doctoral thesis). University of Southampton.

Huang, Y., & Griffin, M. J. (2008). Nonlinear dual-axis biodynamic response of the semi-supine human body during longitudinal horizontal whole-body vibration. *Journal of Sound and Vibration*, *312*(1), 273–295.

Huang, Y., & Griffin, M. J. (2009). Nonlinearity in apparent mass and transmissibility of the supine human body during vertical whole-body vibration. *Journal of Sound and Vibration*, *324*(1), 429–452.

International Organization for Standardization. (1997). *Mechanical vibration and shock—Evaluation of human exposure to whole-body vibration—Part 1: General requirements* (ISO Standard No. 2631-1:1997). https://www.iso.org/standard/7612.html.

Johanning, E., Landsbergis, P., Fischer, S., Christ, E., Göres, B., & Luhrman, R. (2006). Whole-body vibration and ergonomic study of United States railroad locomotives. *Journal of Sound and Vibration*, *298*, 594–600.

Kitazaki, S., & Griffin, M. J. (1998). Resonance behavior of the seated human body and effects of posture. *Journal of Biomechanics, 31*, 143–149.

Kittusamy, N. K., & Buchholz, B. (2004). Whole-body vibration and postural stress among operators of construction equipment: A literature review. *Journal of Safety Research, 35*(3), 255–261.

Liang, C.-C., & Chiang, C.-F. (2008). Modeling of a seated human body exposed to vertical vibrations in various automotive postures. *Industrial Health, 46*(2), 125–137.

Liu, C., & Qiu, Y. (2020). Localised apparent masses over the interface between a seated human body and a soft seat during vertical whole-body vibration. *Journal of Biomechanics, 109*, 109887.

Lundström, R., Holmlund, P., & Lindberg, L. (1998). Absorption of energy during vertical whole-body vibration exposure. *Journal of Biomechanics, 31*(4), 317–326.

Mandapuram, S., Rakheja, S., & Boileau, P.-E. (2015). Energy absorption of seated body exposed to single and three-axis whole body vibration. *Journal of Low Frequency Noise, Vibration and Active Control, 34*(1), 21–38.

Mandapuram, S., Rakheja, S., Boileau, P.-E., Maeda, S., & Shibata, N. (2010). Apparent mass and seat-to-head transmissibility responses of seated occupants under single and dual axis horizontal vibration. *Industrial Health, 48*, 698–714. Available from https://doi.org/10.2486/indhealth.MSWBVI-15.

Mandapuram, S. C. (2012) *Biodynamic responses of the seated occupants to multi-axis whole-body vibration* (PhD thesis). Concordia University.

Mansfield, N. J. (2005a). *Human response to vibration*. Boca Raton, FL: CRC Press.

Mansfield, N. J. (2005b). Impedance methods (apparent mass, driving point mechanical impedance and absorbed power) for assessment of the biomechanical response of the seated person to whole-body vibration. *Industrial Health, 43*(3), 378–389.

Mansfield, N. J., & Griffin, M. J. (2000). Non-linearities in apparent mass and transmissibility during exposure to whole-body vertical vibration. *Journal of Biomechanics, 33*(8), 933–941.

Mansfield, N. J., Holmlund, P., & Lundstrom, R. (2001). Apparent mass and absorbed power during exposure to whole-body vibration and repeated shocks. *Journal of Sound and Vibration, 248*(3), 427–440.

Mansfield, N. J., Holmlund, P., Lundström, R., Lenzuni, P., & Nataletti, P. (2006). Effect of vibration magnitude, vibration spectrum and muscle tension on apparent mass and cross axis transfer functions during whole-body vibration exposure. *Journal of Biomechanics, 39*, 3062–3070.

Mansfield, N. J., & Maeda, S. (2007). The apparent mass of the seated human exposed to single-axis and multi-axis whole-body vibration. *Journal of Biomechanics, 40*(11), 2543–2551.

Matsumoto, Y., & Griffin, M. J. (2002a). Effect of muscle tension on non-linearities in the apparent masses of seated subjects exposed to vertical whole-body vibration. *Journal of Sound and Vibration, 253*, 77–92.

Matsumoto, Y., & Griffin, M. J. (2002b). Non-linear characteristics in the dynamic responses of seated subjects exposed to vertical whole-body vibration. *Journal of Biomechanical Engineering, 124*, 527–532.

McDonald, N., Kriellaars, D., Weldon, E., & Pryce, R. (2020). Head–neck motion in prehospital trauma patients under spinal motion restriction: A pilot study. *Prehospital Emergency Care, 25*(1), 117–124. Available from https://doi.org/10.1080/10903127.2020.1727591.

Meusch, J. C. (2012). *Supine human response and vibration-suppression during whole-body vibration* (MS thesis). The University of Iowa.

Meusch, J. C., & Rahmatalla, S. (2014a). 3D transmissibility and relative transmissibility of immobilized supine humans during transportation. *Journal of Low Frequency Noise, Vibration and Active Control, 33*(2), 125–138. Available from https://doi.org/10.1260/0263-0923.33.2.125.

Meusch, J. C., & Rahmatalla, S. (2014b). Whole-body vibration transmissibility in supine humans: Effect of board, litter, and neck collar. *Applied Ergonomics, 45*(3), 677−685.

Nawayseh, N., Al Sinan, H., Altenejji, S., & Hamdan, S. (2019). Effect of gender on the biodynamic responses to vibration induced by a whole-body vibration training machine. *Proceedings of the Institution of Mechanical Engineers, Part H: Journal of Engineering in Medicine, 233*(3), 383−392.

Nawayseh, N., & Griffin, M. J. (2005). Non-linear dual-axis biodynamic response to fore-and-aft whole-body vibration. *Journal of Sound and Vibration, 282*, 831−862.

Newell, G. S., & Mansfield, N. J. (2008). Evaluation of reaction time performance and subjective workload during whole-body vibration exposure while seated in upright and twisted postures with and without armrests. *International Journal of Industrial Ergonomics, 38*, 499−508.

Newland, D. E. (1984). *An introduction to random vibrations and spectral analysis*. New York: Longman Inc.

Paddan, G. S., & Griffin, M. J. (1988). The transmission of translation seat vibration to the head—I. Vertical seat vibration. *Journal of Biomechanics, 21*, 191−197.

Paddan, G. S., & Griffin, M. J. (1994). Transmission of roll and pitch seat vibration to the head. *Ergonomics, 37*(9), 1513−1531.

Paddan, G. S., & Griffin, M. J. (1998). A review of the transmission of translational seat vibration to the head. *Journal of Sound and Vibration, 215*, 863−882.

Parsons, K. C., & Griffin, M. J. (1978). The effect of the position of the axis of rotation on the discomfort caused by whole-body roll and pitch vibrations of seated persons. *Journal of Sound and Vibration, 58*(1), 127−141.

Qiao, G. (2017). *Identification of physical parameters of biological and mechanical systems under whole-body vibration* (Dissertation). The University of Iowa.

Qiu, Y., & Griffin, M. J. (2004). Transmission of vibration to the backrest of a car seat evaluated with multi-input models. *Journal of Sound and Vibration, 274*, 297−321.

Rahmatalla, S., & DeShaw, J. (2011). Effective seat-to-head transmissibility in whole-body vibration: Effects of posture and arm position. *Journal of Sound and Vibration, 330*, 6277−6286.

Rahmatalla, S., DeShaw, J., & Barazanji, K. (2017). Biodynamics of supine humans and interaction with transport systems during vibration and shocks. *Journal of Low Frequency Noise, Vibration and Active Control, 38*(2), 1−9.

Rahmatalla, S., Kinsler, R., Qiao, G., DeShaw, J., & Mayer, A. (2020). Effect of gender, stature, and body mass on immobilized supine-human response during en route care transport. *Journal of Low Frequency Noise, Vibration and Active Control*. Available from https://doi.org/10.1177/1461348420911253.

Rahmatalla, S., & Liu, Y. (2012). An active head-neck model in whole-body vibration: Vibration magnitude and softening. *Journal of Biomechanics, 45*(6), 925−930.

Rahmatalla, S., Qiao, G., Kinsler, R., DeShaw, J., & Mayer, A. (2021). Stiffening behavior of supine humans during en route care transport. *Vibration, 4*(1), 91−100. Available from https://www.mdpi.com/2571-631X/4/1/8/pdf.

Rehn, B., Nilsson, T., Olofsson, B., & Lundstorm, R. (2005). Whole-body vibration exposure and non-neutral neck postures during occupational use of all-terrain vehicles. *The Annals of Occupational Hygiene, 49*, 267−275.

Shibata, N., Ishimatsu, K., & Maeda, S. (2012). Gender difference in subjective response to whole-body vibration under standing posture. *International Archives of Occupational and Environmental Health, 85*, 171−179.

Smith, D. R., Smith, J. A., & Bowden, D. R. (2008). Transmission characteristics of suspension seats in multi-axis vibration environment. *International Journal of Industrial Ergonomics, 38*, 434–446.

Smith, S. D., & Kazarian, L. E. (1994). The effects of acceleration on the mechanical impedance response of a primate model exposed to sinusoidal vibration. *Annals of Biomedical Engineering, 22*, 78–87.

Thuresson, M., Ang, B., Linder, J., & Harms-Ringdahl, K. (2005). Mechanical load and EMG activity in the neck induced by different head-worn equipment and neck postures. *International Journal of Industrial Ergonomics, 35*, 13–18.

Toward, M. G., & Griffin, M. J. (2011). Apparent mass of the human body in the vertical direction: Inter-subject variability. *Journal of Sound and Vibration, 330*(4), 827–841.

Vogt, H. L., Coermann, R. R., & Fust, H. D. (1968). Mechanical impedance of the sitting human under sustained acceleration. *Aerospace Medicine, 39*, 675–679.

Vogt, L. H., Krause, H. E., Hohlweck, H., & May, E. (1973). Mechanical impedance of supine humans under sustained acceleration. *Aerospace Medicine, 44*(2), 123–128.

Wang, S. C., et al. (2004). Gender differences in hip anatomy: Possible implications for injury tolerance in frontal collisions. *Annual Proceedings/Association for the Advancement of Automotive Medicine, 48*, 287–301.

Wang, W., Rakheja, S., & Boileau, P. (2006). The role of seat geometry and posture on the mechanical energy absorption characteristics of seated occupants under vertical vibration. *International Journal of Industrial Ergonomics, 36*, 171–184.

Wang, Y., & Rahmatalla, S. (2013). Human head–neck models in whole-body vibration: Effect of posture. *Journal of Biomechanics, 46*(4), 702–710.

Wei, L., & Griffin, M. (1998). Mathematical models for the apparent mass of the seated human body exposed to vertical vibration. *Journal of Sound and Vibration, 212*(5), 855–874.

Chapter 4

Discomfort in whole-body vibration

4.1 Introduction

Humans can be subjected to different types of discomfort during their daily activities. Static discomfort and quasistatic discomfort, which result from holding stationary postures or conducting tasks with relatively slow movements, have been recognized for a long time. Sitting in a static position for a long time and assuming awkward postures can cause severe discomfort and lead to chronic back and neck pain and the potential for nerve damage. Lying in a bed for medical reasons for a prolonged time can cause pressure sores and different types of musculoskeletal and neuromuscular problems. Dynamic discomfort, on the other hand, occurs when people are exposed to vibration. People may have noticed dynamic discomfort with relatively severe motions during ancient times, when they started riding horses and using carts for transportation and battles. Dynamic discomfort became more obvious with the start of the Industrial Revolution and the invention of faster transport systems and the production of heavy machinery and tools with which vibration can be significant. Since then, more advanced transport systems with appropriate comfort levels started to appear on the market. The situation became more challenging when people started working in environments while subjected to whole-body vibration (WBV) such as those in the railroad, construction, agriculture, and mining industries. The biodynamic response of the human body under such dynamic conditions can lead to excessive voluntary and involuntary motions that can cause additional discomfort that can be added to the static discomfort and lead to fatigue and potential harm. This is critical for medical transport when both medics and wounded patients can be exposed to harm. Therefore the design of transport systems for WBV applications should consider both the static and dynamic components of discomfort.

Much work has been done in the literature to understand and quantify discomfort in static and dynamic conditions for seated and standing positions; in comparison, less work has been done on supine positions. Moreover, discomfort measurement and quantification are traditionally conducted using subjective measures that are based on the reported discomfort

Prehospital Transport and Whole-Body Vibration. DOI: https://doi.org/10.1016/B978-0-323-90103-1.00004-8
141

of the human subjects using questionnaires. Methods for objective evaluation of discomfort have also evolved with time. Approaches to objectively quantifying static discomfort are mostly based on the measurement of the stiffness and the pressure distribution on the interface between the contact surfaces and the human body. Methods for evaluating dynamic discomfort are more complicated as it involves many factors, such as the magnitude, frequency content, and direction of vibration. It is also sensitive to people's posture and the way they engage with the surrounding equipment, among other things. Many methods have been developed to quantify dynamic discomfort, and most use metrics that are based on the input vibration at the interface between the supporting surfaces and the human body during vibration. Little attention has been paid, in these dynamic approaches, to the measurement and the role of the relative motion of the human body segments during vibration. Also, most existing practices for dynamic discomfort evaluation are based on the translational motions during vibration, while overlooking the role of the angular motions. This chapter presents some ongoing and new approaches that evaluate human discomfort in WBV with some emphasis on the role of angular motion. The chapter also introduces mathematical and statistical schemes that can be used to quantify and predict discomfort during WBV.

4.2 Methods of discomfort quantification

Static discomfort has been widely investigated in the literature and has been used to advance the comfort of seats and transport systems. The process of quantifying static discomfort is mostly based on subjective or reported evaluations using questionnaires when humans are sitting or lying on these systems. More advanced methods for assessing static discomfort have evolved using objective measurements such as the stiffness and the pressure distribution at the interface between the contact surfaces and the human body. The application of these objective metrics has shown promising results for the design of seats for offices and the car industry (Kolich, 2004). Yet, the use of static discomfort to evaluate systems in vibration environments has shown drawbacks and limitations in many applications. One example is the bad rating of seats that showed good static discomfort but did not provide adequate comfort during WBV. Another example is the ineffectiveness of patients' transport systems, which are evaluated based on static discomfort, in providing adequate comfort in a WBV environment.

4.2.1 Dynamic discomfort—history

Early studies of dynamic discomfort during WBV and supine-human transport go as far back as 1931, when Reiher and Meister (1931) conducted studies quantifying vibration displacement and comparing it to perception and feelings to vibration. Fast forward to our current times, and with recent advancements in transportation technologies, new challenges are arising, such as those in high-speed trains, when people can be subjected to relatively high vibration and shock in the vertical and lateral directions (Johanning et al., 2002), and which can lead to discomfort (Lee et al., 2009); this is especially true for people using sleeping berths during transport (Peng et al., 2009) or during the speedy transport of critically injured patients during wars and natural disasters. Quantification of dynamic discomfort in WBV environments has been studied by many researchers (DeShaw & Rahmatalla, 2016; Dickey et al., 2006; Ebe & Griffin, 2000b; Mansfield & Maeda, 2011; Miwa, 1967) with more emphasis on seated positions (Zhou & Griffin, 2014; Hacaambwa & Giacomin, 2007; Mansfield et al., 2014). More works have been conducted recently to study discomfort in standing and supine positions such as those in railways (Liu et al., 2019; Peng et al., 2009) and medical transport (Goswami et al., 2020).

4.2.2   Subjective evaluation of discomfort

Discomfort is traditionally evaluated during WBV based on subjective reported measures (Hacaambwa & Giacomin, 2007; Kaneko et al., 2005; Mansfield & Maeda, 2011). Many factors can affect the reported discomfort in WBV, including the vibration magnitude, the frequency content of vibration, the vibration direction, the posture, the quality of contact surface, and the way the humans interact with the surrounding equipment. The interpretation of the subjective or reported discomfort in WBV studies can sometimes be difficult, as people rate their discomfort differently (Hwang & Yoon, 1981); some people give high ratings, and some give low ratings when tested in the same scenario. Therefore normalization methodologies are normally used to standardize the reported discomfort between subjects (Hwang & Yoon, 1981).

In WBV, the reported discomfort includes static discomfort, which represents the discomfort level of the subjects before vibration is applied, and dynamic discomfort, which includes discomfort resulting from vibration as the body shakes around. Some studies start the process of evaluating discomfort by reporting the discomfort of the subjects before vibration and using that as a baseline that will be added to the total discomfort coming from vibration and other means (DeShaw & Rahmatalla, 2016; Mansfield et al., 2014). Methods to report subjective discomfort are generally based on questionnaires about how subjects rate their discomfort under certain conditions. The Borg CR-10 scale (Fig. 4.1) (Borg, 1982) is a very popular measure that is traditionally used in pain and discomfort evaluations. The scale ranges from 0 to 10, with 0 indicating no discomfort and 10 indicating the most severe discomfort. For the supine position, subjects at a 0 on the Borg scale may be considered to be lying comfortably in a bed.

Borg CR-10 Scale	
0	Nothing at all
0.3	
0.5	Extremely weak
0.7	
1	Very weak
1.5	
2	Weak
2.5	
3	Moderate
4	
5	Strong
6	
7	Very strong
8	
9	
10	Extremely strong
*	Absolute maximum

FIGURE 4.1   Borg CR-10 scale.

4.2.2.1   Analysis of reported discomfort data

Discomfort reported by subjects is traditionally normalized to give a value between 0 and 1. This process is done to minimize the differences between the reported discomfort of different subjects, as there is a tendency for some subjects to report higher or smaller numbers than other subjects. The method outlined by Hwang and Yoon (1981) normalized reported discomfort as follows:

$$ND = \frac{RD - Min(RD \text{ of subject})}{(Max(RD \text{ of subject}) - Min(RD \text{ of subject}))} \qquad (4.1)$$

where ND is the normalized discomfort and RD is the reported discomfort. The ND data can be transformed and rescaled into the Borg CR-10 scale using the approach by DeShaw and Rahmatalla (2016), as follows:

$$ND_{scaled} = \frac{RD - Min(RD \text{ of subject})}{(Max(RD \text{ of subject}) - Min(RD \text{ of subject}))} *Avg \text{ Max of all subjects}$$

$$(4.2)$$

4.2.2.2   Case study: subjective evaluation of discomfort

Twelve male subjects with no history of neck, shoulder, or head injuries, or any neurological conditions, participated in this study (DeShaw &

Rahmatalla, 2016). The study involved exposing the subjects to different types of WBV while they were lying on transport systems with different supporting surfaces and immobilization conditions (Fig. 4.2). The transport systems were rigidly attached to a man-rated shaker table, which allows safe testing of humans and can simulate different types of vibration scenarios in WBV. The subjects were exposed to WBV in the three translational directions (vertical in the *X*-direction, fore-aft in the *Z*-direction, and side-to-side in the *Y*-direction). The vibration files were also mixed with shock signals. The ride files contained random vibration files with a frequency content of

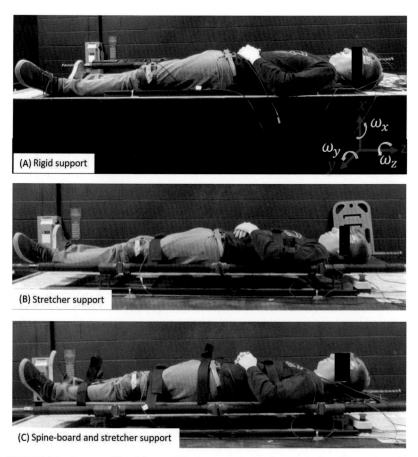

FIGURE 4.2 Human subject lying on different supporting surfaces and using different immobilization or supporting conditions: (A) rigid support, (B) stretcher support, and (C) spinal backboard support and stretcher. The small *orange* boxes on the pelvis and knee are inertial sensors. *Adapted from the work of DeShaw, J., & Rahmatalla, S. (2016). Predictive discomfort of supine humans in whole-body vibration and shock environments.* Ergonomics, 59(4), 568–581. *https://doi.org/10.1080/00140139.2015.1083125.*

0.5−25 Hz. Inertial sensors were attached at the interface between the human body and the supporting surfaces near the middle of the supporting surfaces and under the pelvis area. Sensors were also attached to the human body at the forehead just between the eyebrows, the sternum, the left anterior superior iliac spine, and the patella of the left knee, as shown in Fig. 4.2.

4.2.2.2.1   Data collection

At the beginning of the study, before the vibration was applied and while the subjects were in their static positions, they were asked to rate their static discomfort based on the Borg CR-10 scale. The subjects were also allowed to report discomfort at any other region and the intensity or maximum intensity of discomfort at these regions. Then the subjects were exposed to random WBV of different magnitudes and directions, including three-directional (3D) input (3IP) with high and low magnitude (3D high and 3D low) and 3IP with high and low magnitude mixed with shocks (S) (3D + S high and 3D + S low). Additional vibration rides with different magnitudes were used for validation purposes and included files with the letter T such as 3IP (3D T1 and 3D + S T2). The length of each ride file was 60 seconds. Based on the literature (Dickey et al., 2006; Miwa & Yonekawa, 1969), a 15-second duration of vibration was considered appropriate for the subjects to feel and report their discomfort value. Subjects were asked to consider their discomfort at 45 seconds and then to finalize and report their discomfort at the end of the 60 seconds. The subjects were also asked to localize the areas of discomfort on their body during the ride file. Additionally, the subjects were asked to report on any particular areas of discomfort during the WBV.

4.2.2.2.2   Analysis of the data

The collected reported discomfort data were first normalized using Eq. (4.1) and then transformed to the Borg CR-10 scale using Eq. (4.2). Fig. 4.3 shows the average normalized static and dynamic discomfort for the 12 subjects under different supporting systems and vibration conditions. The lower part of each bar represents the static discomfort portion of the perceived discomfort, while the upper part represents the reported dynamic discomfort as a result of vibration. It is clear from the figure that the dynamic component of discomfort can be multiple times larger than the static components of discomfort. While the static discomfort depends on the stiffness of the supporting surfaces and the pressure distribution, the rigid support has some of the highest values, followed by the board configuration. The dynamic discomfort, on the other hand, depends on the severity of motion, and therefore the highest dynamic discomfort is associated with the 3D files mixed with shocks.

The bar graph in Fig. 4.4 demonstrates the average percentage of discomfort across subjects' bodies under the different vibration exposure and

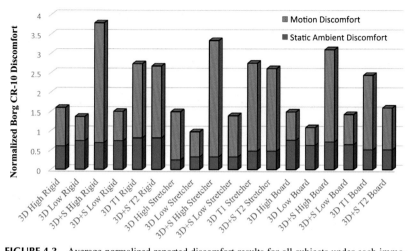

FIGURE 4.3   Average normalized reported discomfort results for all subjects under each immobilization and vibration condition. The static component of the discomfort is indicated by the lower *blue* portion of the bars, and the dynamic/motion component of discomfort is indicated by the upper *gray* portion. Data is from *DeShaw, J., & Rahmatalla, S. (2016). Predictive discomfort of supine humans in whole-body vibration and shock environments.* Ergonomics, *59(4), 568−581. https://doi.org/10.1080/00140139.2015.1083125.*

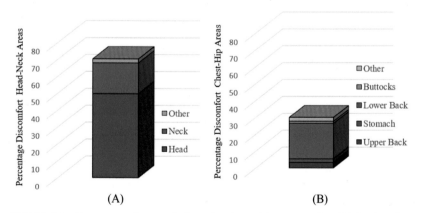

FIGURE 4.4   Percentages of reported discomfort lumped into categories: (A) head-to-chest areas, and (B) chest-to-pelvis areas. Each slice in the figure represents the magnitude of discomfort at the different regions of the human body. Data is from *DeShaw, J., & Rahmatalla, S. (2016). Predictive discomfort of supine humans in whole-body vibration and shock environments.* Ergonomics, *59(4), 568−581. https://doi.org/10.1080/00140139.2015.1083125.*

immobilization conditions. The value of discomfort at the critical locations on the human body is represented by slices. The thickness of each slice represents the relative magnitude of discomfort at these locations. The data analysis showed that the majority of discomfort was located in the head, neck, and lower back areas. Fig. 4.4A shows that discomfort at the head and neck areas comprises the majority of the perceived discomfort under the vibration scenarios in this case study. Fig. 4.4B shows that the intensity of discomfort at the lower back area is greater than at other regions.

4.2.3 Objective evaluation of discomfort

Much work has been done to develop objective measures to evaluate or predict discomfort during WBV. The most prominent and well-accepted approach is the one that is outlined by standards such as ISO 2631-1 (International Organization for Standardization, 1997) and BS 6841 (British Standards Institution, 1987). Other objective measures have been presented in the literature but are considered to be at the research level. These include the use of transfer functions, the power law, equal sensations and comfort contours, and predictive measures that consider the rotational motions of human body segments.

4.2.3.1 ISO evaluation of discomfort

The early work in the ISO standard committees to develop metrics for quantifying objective discomfort in WBV may have started in the mid-1960s, but the first version of the standard ISO 2631-1 for determining comfort during WBV was established in 1974 (Oborne, 1983). The current reference standards are ISO 2631-1 Mechanical vibration and shock—evaluation of human exposure to WBV (International Organization for Standardization, 1997) and the British standard BS 6841 (British Standards Institution, 1987). These two standards are similar in many ways, and readers are referred to the work of Mansfield (2005) and Lewis and Griffin (1998) for specific comparisons. Both standards use acceleration for the quantification of human perception, comfort, and health under WBV. While the latter parameters can be affected by vibration magnitude, the standard uses the magnitude of the input accelerations, normally the root mean square (RMS) values, at the interface between the human body and contact surfaces such as the seat pan, seat backrest, and footrest for seated humans and the supporting surfaces of the transport system for supine humans as a means for quantifying discomfort. Depending on the type of supporting surface for supine humans, the input acceleration measurements should be conducted at the pelvis, back, and/or head. ISO 2631-1 also considers the effect of the rotational input motions on human comfort. The total value of the input rotational motions may be included in the RMS calculations when determining the overall vibration, using equations similar to those used for translational vibrations. ISO 2631-1 also describes methods that include rotational vibration and vibration at the feet and seat back (Griffin, 1990; Lewis & Griffin, 1998; Marjanen & Mansfield, 2010).

Because human response and perception to vibration are affected by the frequency content of the vibration signal, the input acceleration values are usually weighted with respect to the frequency to develop frequency-weighted accelerations. The standard uses these frequency-weighted accelerations to create different metrics and develop guidance that relates the different metrics to fatigue and comfort limits. During the frequency-weighting process, the frequencies to which the human body is less sensitive to vibration are given lower weights than the frequencies to which the human body is very sensitive; meanwhile, the frequency weighting is designed such that they do not affect the frequencies at which the human response is sensitive (Mansfield, 2005). For example, when a seated human is exposed to sinusoidal vertical vibration, the human may not generate much discomfort at low frequencies of 1−2 Hz or at high frequencies of 40 Hz; however, because of the resonance phenomenon, the human body can generate excessive motions when exposed to frequencies between 4 and 8 Hz. In this case, the frequencies at 1−2 and 40 Hz will be given less weightage and the frequencies between 4 and 8 Hz will not much be affected by weighting. The current ISO standard considers vibrations occurring between 1 and 80 Hz as frequencies of interest for exposure of humans to WBV.

4.2.3.1.1   Weighted RMS acceleration evaluation

The standard uses the following equation to calculate the RMS frequency-weighted acceleration (a_ω) entering the human body at the interface with the contact surfaces.

$$a_\omega = \sqrt{\frac{1}{T}\int_0^T a_\omega^2(t)dt} \qquad (4.3)$$

where $a_\omega(t)$ is the frequency-weighted acceleration as a function of time (t) in seconds. Frequency weighting are designed to not affect the acceleration magnitude at frequencies where the body is sensitive to vibration. So the weighted acceleration should be always less than the unweighted acceleration (Mansfield, 2005). This is because weighing uses 1 (100%) for the highest value and all the rest will be multiplied by a fraction of 1. This is how all weighted values end up same or smaller than the unweighted ones. The frequency-weighted acceleration can be also calculated in the frequency domain using an equivalent form of the frequency-weighted acceleration using one-third octave band of unweighted acceleration. The one-third octave band is a band of frequency where the upper band-edge frequency is equal to the lower band frequency times the cube root of two. The one-third octave ban is normally used to avoid using the whole frequency range in the analysis. In this case, each frequency in the center of each one-third octave band frequency will be multiplied (weighted) by a certain value that can be obtained in tables or graphs outlined in the ISO standard.

The formula for this transformation is as follows:

$$a_\omega = \left[\sum_i (W_i a_i) \right]^{1/2} \qquad (4.4)$$

where a_ω is the resulting weighted acceleration, Wi is a weighting factor specified in the ISO 2631-1, and a_i is the RMS value of unweighted acceleration in the ith one-third octave band.

4.2.3.1.2   Effect of vibration direction

In addition to the effect of frequency content on human comfort in WBV, humans are also sensitive to the direction of vibration. For seat comfort evaluation, ISO-2631-1 includes the effect of input vibration in 12 axes of vibration including 3 translational axes at the floor, 3 rotational axes and 3 translational axes at the seat-cushion interface, and 3 translational axes at the seatback (Marjanen & Mansfield, 2010). However, most studies ignore the rotational components, seatback components, and floor and use only three translational axes at the seat-cushion interface. For recumbent persons, three input translational directions are normally conducted at the interface with the contact surfaces beneath the pelvis, but the measurement can be extended to include the areas beneath the back and head. The standard uses weighting factors for each direction. The values of these weighting factors were based on experimentation and expert opinions. The weighting factors for a recumbent person when the input accelerations are measured under the pelvis are W_d, $k = 1$ for horizontal axes and W_k, $k = 1$ for the vertical axis. When there is no cushion or padding under the head of a supine person, the standard recommends measuring the acceleration under the head and using a factor W_j with $k = 1$. In order to assess a vibration that takes place in more than one direction simultaneously, ISO 2631-1 suggests that the effect of such a vibration can be calculated by taking the vector sum, a_V, of the three weighted acceleration values as follows:

$$a_V = (k_x^2 a_{\omega x}^2 + k_y^2 a_{\omega y}^2 + k_z^2 a_{\omega z}^2)^{1/2} \qquad (4.5)$$

where $a_{\omega x}$, $a_{\omega y}$, and $a_{\omega z}$ are the weighted RMS acceleration with respect to the x-, y-, and z-directions, and k_x, k_y, and k_z are the corresponding multiplying factors. Values for k_x, k_y, and k_z are specified to be 1.4, 1.4, and 1, respectively, for seated subjects. The origin of the 1.4 factor was selected based on experts' opinions through discussion in the ISO committees. The standard may also consider the perception of vibration as the highest weighted RMS acceleration in any axis at the input contact point at any time (Mansfield, 2005). In their studies on the relative contribution of the vibrational axes to discomfort for 22 seated male and female subjects, Marjanen and Mansfield (2010) presented optimized multiplication factors for k_x, k_y, and k_z to be 2.7, 1.8, and 1, respectively, which means higher weights for the fore-aft and lateral directions.

When excessive shocks are presented in the vibration rides, these shocks can generate severe discomfort and potential harm to the human body. Under such circumstances, the RMS values of the input acceleration may underestimate the severity of the vibration entering the body. Therefore the standard presents more effective metrics to deal with such conditions. One of these metrics is the crest factor, which is the ratio between the maximum peak of the shocks in the acceleration signal divided by the RMS value of the signal. The peak should be determined based on the time period used for the integration of the RMS value, as follows:

$$CF = \frac{\max(a_\omega(t))}{a_\omega} \qquad (4.6)$$

where CF is the crest factor, max $(a_\omega(t))$ is the maximum instantaneous peak value of the RMS acceleration at time t, and $a_\omega(t)$ is the frequency-weighted RMS acceleration. Acceleration signals with a crest factor higher than 9 should be addressed using the approaches outlined in ISO 2631-5 (International Organization for Standardization, 2004) or other methods outlined in ISO-26331-1 (International Organization for Standardization, 1997). These methods include using the vibration dose value (VDV) and the maximum transient vibration value (MTVV) of the frequency-weighted acceleration to deal with acceleration signals that have occasional peaks (shocks). The VDV formula, shown in Eq. (4.7), is to the power 4, which makes it very sensitive to peaks in comparison with other acceleration values.

$$VDV = \sqrt[4]{\int_0^T a_\omega^4(t)dt} \qquad (4.7)$$

where $a_\omega(t)$ is the frequency-weighted RMS acceleration at time t and T is the duration of the measurement.

The MTVV is another metric used in ISO-2631-1 to deal with acceleration signals containing shocks. The MTVV is calculated as follows:

$$MTVV = \max[a_\omega(t_0)] \qquad (4.8)$$

where $a_\omega(t_0)$ is the highest magnitude during the measurement period (T). T is recommended to be 1 second.

The RMS and VDV of the frequency-weighted acceleration as defined in ISO 2631-1 (International Organization for Standardization, 1997) and BS 6841 (British Standards Institution, 1987) are the most common methods for evaluating vibration and predicting discomfort when subjects are exposed to vibration. However, Ebe and Griffin (2000a) reported that such measures will not always correlate with comfort evaluations, especially when the magnitude of vibration is low.

The standards in general do not present metrics to quantify discomfort, so no limits are defined; however, ISO-2631-1 suggests that vibration values obtained from one environment may be compared to those obtained in another environment so as to compare the discomfort. While the standard emphasizes that discomfort can be affected by many factors, including people's activity and perception when exposed to WBV, it suggests some general values for guidance. For example, an input weighted RMS acceleration of less than 0.315 m/s^2 is not considered uncomfortable, while a weighted RMS acceleration greater than 2 m/s^2 is considered extremely uncomfortable. Also, as guidance for vibration perception, which is considered to vary widely among individuals, the standards state that the interquartile range may extend from 0.01 to 0.02 m/s^2.

4.2.3.1.3   Validity and limitations of the current standards

The ISO 2631-1 standard gives detailed information about the measurement and evaluation of health, discomfort, and vibration perception for seated, standing, and recumbent operators during WBV. For the recumbent or supine postures, however, the standards give only general guidance about discomfort. Since its establishment, ISO 2631-1 has undergone many evaluations and discussions that have shown the limitations of the standard in different applications. For example, the current standards consider the input acceleration as metrics for discomfort evaluations; however, they do not include the effect of the comfort due to static discomfort, which can be affected by the quality of the seat or the contact surfaces used in the testing environments. The latter could be very important, especially when the vibration is taking place at low frequencies (Ebe & Griffin, 2000a). In this case, the stiffness of the contact surface may have more effect on discomfort than vibration (Ebe & Griffin, 2000a). Another issue with the standard is that it does not consider the effects of different postures that humans take during their daily work activities. This could be critical when evaluating environments in which a person is required to take awkward postures while they are subjected to WBV. One example is medics or emergency responders when they are supporting the lives of injured patients inside a helicopter or a Humvee and are forced to take awkward postures for prolonged times. Much work has shown the limitations of the current ISO 26331-1 to deal with human postures taken during vibration. For example, Basri and Griffin (2013) showed that there is no clear requirement in either the frequency weightings or the axis multiplication factors to consider the differences in the inclination angle of the backrests. Studies were also conducted to investigate the effect of multiaxis vibrations on discomfort. Some works showed discomfort values that differed from what the standard predicts. Marjanen and Mansfield (2010) conducted a laboratory study to validate the methods used in the ISO 2631-1 standard when dealing with multiaxis vibration. The study revealed that the multiplying factors in the standards, which are used to give weights to each direction, are not optimal. The authors suggested better values for the multiplying factors for all measured axes. There are also ongoing debates on the validity of ISO 2631-1 for several applications. Kaneko et al. (2005) showed that subjects felt different levels of discomfort when they were subjected to random vibration stimuli made up of different frequency ranges, while the frequency-weighted RMS acceleration by the ISO-2631-1 predicts the same level of discomfort.

On the other hand, there were positive signs about the capability of the standards to evaluate discomfort in complicated environments. Fairley and Griffin (1988) investigated the prediction of discomfort caused by simultaneous exposure to vertical and fore-aft WBV. In this process, the seated subjects were asked to adjust the combinations of vertical and fore-aft whole-body sinusoidal vibration with a frequency range of 2−10 Hz so their discomfort would be equivalent to the discomfort of a reference motion. It was concluded that the RMS procedure outlined in ISO 2631-1 and BSI 6841 is the most appropriate method for predicting the discomfort of WBV with combined vertical and fore-aft vibration. In their investigation about the effectiveness of ISO 2631-1 in predicting discomfort for multiaxis vibration with seated subjects taking different postures, DeShaw and Rahmatalla (2014) demonstrated the limitations of the current standards in dealing with postures in WBV and the need to improve them to accommodate postures in discomfort evaluation. However, the study did show that the standard produced realistic predictions in cases that did not include posture.

4.2.4   Approaches for evaluation of objective discomfort

Human sensation to vibration was explored a long time ago by many researchers, including Reiher and Meister (1931), who assessed human sensation to vibration in standing and lying postures. Miwa (1967) evaluated the threshold of vibration perception and used the equal sensation contours of sitting and standing people during horizontal and vertical vibration. The equal sensation contour represents values of equal subjective intensity plotted with vibration intensity and frequency graphs (Oborne, 1983). The threshold of vibration perception refers to the lowest magnitudes of vibration that subjects can feel at different frequencies. It can also be considered as the vibration magnitude with the frequency at which subjects stop feeling the vibration. This threshold can be determined by exposing the subjects to high vibration values and then reducing the vibration magnitude until the subjects no longer feel the vibration. This can be also verified by exposing the subjects to low vibration that increases in magnitude until they feel the vibration again. The equal sensation curves are constructed by drawing the vibration magnitude with frequencies for different cases of sensation. These studies normally used a reference vibration at a certain magnitude and frequency.

For example, Miwa used a reference vibration with a frequency of 20 Hz at a vibration acceleration level (VAL) value of 20. The VAL represents the ratio between the applied acceleration and a reference acceleration that is selected by the tester. The subjects, while being exposed to vibration at a certain frequency, must tell the tester when they feel the same sensation as that of the reference excitation. This process is repeated for each frequency until all frequencies have been covered. In this case, all points on the equal sensation curve will give sensations equal to the standard reference vibration at 20 Hz with 20 VAL.

While it is known from experimentations that discomfort can be affected by vibration magnitudes, people have always been interested in finding mathematical forms that resemble these relationships. During the early work, researchers tried to find a relationship between how people perceived vibration and felt discomfort based on concepts from sounds and acoustics. One popular form is the power law, in which the relationship between the input and output is expressed in exponential or power forms. One of the earliest approaches was Stevens' psychophysical law (Stevens, 1957), in which the relationship between the stimulate (vibration magnitude) and the sensation (perceived discomfort magnitude) is presented in the form of a power law:

$$P = kV^n \tag{4.9}$$

where P represents a sensation magnitude and V is the stimulus magnitude. The parameters n and k are constants that can be found from experiments using a regression form of the formula. In this process, Eq. (4.9) is transformed to the following form:

$$\log P = \log K + n \times \log V \tag{4.10}$$

The values of n and k can be determined from the regression line's slope and its intersection with the vertical axis, respectively. As shown in Fig. 4.5, n is the slope of the line and k can be calculated from the intersection with the vertical axis.

The power law relates discomfort with the vibration magnitude; after calculating the constants of the formula (n and k), a relationship between vibration magnitude and discomfort can be established. Because humans are sensitive to vibrations with different frequencies, the parameter (n) in the power law should not be constant but rather be expected to change with the frequency. Therefore a variety of efforts were made to include the effect of the frequency in the power law. One effort was to present the discomfort as a function of two variables, vibration magnitude and frequency: the discomfort will become a two-dimensional surface as a function of the vibration magnitude and frequency. Slices of the discomfort surface, where discomfort has a constant value, called comfort contours, can then be generated (Morioka & Griffin, 2006). In this way, each discomfort contour line will show how the vibration magnitude is changing with frequencies to give the

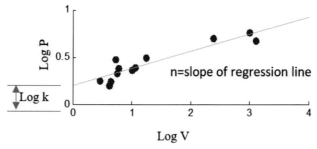

FIGURE 4.5 Estimating the values of the power law coefficients (n and k) from the line regression formula. *Adapted from DeShaw, J., & Rahmatalla, S. (2016). Predictive discomfort of supine humans in whole-body vibration and shock environments.* Ergonomics, *59(4), 568–581. https://doi.org/10.1080/00140139.2015.1083125.*

same discomfort or sensation. Because the vibration magnitude is changing with frequency to generate the same discomfort at different frequencies, the exponential factor (n) in the power law will not be constant but will be a function of the frequency.

Zhou and Griffin (2014) investigated the equivalent comfort contours of the vertical WBV of 20 male and 20 female seated subjects using frequencies of $1-16$ Hz. In their study, the subjects were exposed to a reference vibration signal with a frequency of 4 Hz at one of three magnitudes (0.125, 0.315, or 0.8 m/s^2 RMS), with a constant magnitude within a session. During the experiments, the subjects were asked to report their discomfort level as compared to the reference vibration. When the subjects perceived the vibration to be the same as the reference signal, the discomfort value was registered as 100. Discomfort was registered as 50 or 200 if the sensation was half or double that of the reference vibration, respectively. Using the regression form of the power law (Eq. (4.10)), the rate (n) in the power law equation (Eq. (4.9)) was determined for each frequency. The study found that n was highly dependent on the frequency of vibration and can be represented as a function of frequency. Morioka and Griffin (2006) conducted a study on 12 seated subjects under a frequency range of $2-315$ Hz in each of the three orthogonal axes (vertical, fore-aft, and side-to-side) and showed that the exponent n was highly dependent on the frequency and direction of vibration. Similar work was conducted to study the characteristics of the rate of discomfort growth (n) under different vibration directions (Arnold & Griffin, 2018). It should be mentioned that the reference vibration used for comparison was different for different studies. For example, Arnold and Griffin (2018) used 0.25 m/s^2 RMS at 3.15 Hz (frequency-weighted), and Morioka and Griffin (2006) used a magnitude of 0.5 m/s^2 RMS for the vertical axis and 1.0 m/s^2 RMS for lateral axes at a frequency of 20 Hz.

Work was also done to include the effect of seated postures in the power law calculations. In their studies on the effect of seat inclination angles on the discomfort of 12 seated subjects using equivalent comfort contours, Basri and Griffin (2013) showed how the subjects' discomfort depended on the vibration frequency and the existence of a backrest with different inclination angles.

Arnold and Griffin (2018) also investigated the effect of vibration direction on discomfort using the power law. They explained that people perceived a difference in their discomfort when exposed to vibration with different frequencies and directions because these different forms of vibration can cause discomfort in different parts of the body. For example, with high-magnitude vertical vibration, the head, neck, and shoulders were dominant sources of discomfort at lower frequencies (1.0–4 Hz) but not at higher frequencies. Normally, in such studies, researchers used the concept of equivalent comfort contours. Ebe and Griffin (2000a) explored the idea of modifying Stevens' psychophysical law (Stevens, 1957) and developing a predictive discomfort measure. The proposed predictive discomfort measure comprised two components: the first is associated with seat stiffness, which reflects the static discomfort, and the second is associated with the vibration magnitude at the interface between the subjects and the cushions they are sitting on, which represents the dynamic discomfort. When compared with the reported discomfort, they found that their proposed discomfort function had better prediction than the one depending only on vibration magnitude. Fairley and Griffin (1988) also worked on extending a certain form of the power law to two-directional vibration with the goal of determining the best way to predict discomfort caused by simultaneous exposure to vertical and fore-aft vibration. In addition, Mansfield et al. (2014) extended the concept of discomfort by adding fatigue discomfort as another important component, especially when humans are exposed to prolonged WBV. The proposed objective function included a static component dependent on the contact surface stiffness properties and a dynamic component resulting from exposure to WBV. They found that discomfort accumulates with time with a static posture and that subjects experienced more discomfort if they were also exposed to WBV. In other words, exposure to WBV accelerates the development of discomfort. The authors developed a multifactorial model to describe the relationship between the discomfort components, including the reported discomfort, dynamic discomfort, and fatigue-related discomfort that relates to the exposure time.

4.2.4.1 Predictive discomfort using transfer functions

The idea of using transfer functions as predictive measures of discomfort has been explored by many researchers. This was based on the concept that transfer functions represent the energy entering or passing through the body, which could generate discomfort depending on the vibration magnitude and frequency content. Also, the graphs of the transfer functions normally show the characteristic of human response to vibration and at which frequency humans show the highest responses. These transfer functions can be used as weighting curves for weighting the input acceleration to the human body with respect to the frequencies. To this end, Mansfield et al. (2000) used absorbed power, Mansfield and Maeda (2007) used apparent mass, Holmlund et al. (2000) used driving point mechanical impedance, and Paddan and Griffin (1994) used transmissibility. Most of these transfer functions showed good correlations with the reported discomfort. Still, these functions can capture the global response of the human body but may overlook the local response of the individual body segments, especially where there are interactions between the body segments and the surrounding equipment. While most transfer functions are affected by human postures, it is hard to infer or interpret information about how that effect is related to discomfort. For example, Mansfield and Maeda (2005) found that predictive discomfort based on apparent mass will have difficulty predicting the peak in discomfort when the subjects twist their torsos or lift their arms during vibration. Additionally, although transfer functions can capture the effect of vibration direction, they can have complex forms and can generate many graphs that are hard to interpret, especially when the motion involves translational and rotational components.

Most existing predictive discomfort measures use the input motion or forces to the human body and overlook the motion of individual body segments during vibration. It will be shown later in this chapter that the inclusion of body segment motion facilitates the inclusion of posture effects. Posture can play a major role in discomfort when operators hold postures for a long time while working in environments where WBV exists.

4.2.4.2   Dynamic discomfort and the role of human-segments motion

During the evaluation of discomfort and health in WBV, the measurements of the input motion and forces are traditionally conducted at the interface between the human body and the contact surfaces, with little consideration to what happens to the motion of the human and their different body segments. This perception is based on the idea that the magnitude of vibration entering the human body is the main cause of what will happen to the human body when exposed to vibration. While this approach can capture the global motion of the whole body, it does not have the capability to capture the local motions between body segments. Relative motion between body segments is very relevant to discomfort; for example, everyone can feel the discomfort when the head and other body segments start moving and shaking relative to each other and generating involuntary motions during exposure to severe vibration, especially when the vibration generates local effects and excites the natural frequencies of some segments of the human body. The following case study will shed some light on the role of body segment motion on perceived discomfort during WBV.

4.2.4.3   Case study: role of body segments motion

This case study demonstrates the importance of considering body segment motions in the evaluation of discomfort in WBV (Rahmatalla & DeShaw, 2011b). The study was initiated to investigate the reasons behind the low discomfort rating of a new seat that was developed for use in heavy construction machinery. The seat was ergonomically developed in labs to provide a high level of static and dynamic comfort, and it was deemed comfortable when tested according to the ISO 2631-1 standard. When the seat was tested in the field, however, the developers were surprised to discover that machine operators gave the seat a low comfort rating and preferred the old seats over the new, state-of-the-art seat. In order to solve this mystery, the developers decided to conduct controlled investigations inside a lab where real-life applications could be simulated. Both the new and old seats were attached to a human-rated shaking table that can simulate ride files from the field. A screen was installed on the wall facing the subjects, creating virtual reality scenarios in which field operations could be seen and executed (Fig. 4.6D). The subjects, who were professional construction machine operators, were able to communicate with the graphical simulations via control systems attached to their seats or the steering wheel (ST) in front of them. This setup allowed the operators to conduct simulated real-life operations while they were under vibration. The goal of using the virtual reality system was to put the subjects in a semireal environment so they would focus on their tasks and not on the vibration; this was expected to reduce bias while rating discomfort level.

Two seat configurations simulating the old environment were tested beside the new seat configuration, as shown in Fig. 4.6. Fig. 4.6A shows one of the old seating environments, which comprised a seat from which the operator controlled the machine using a ST. Fig. 4.6B shows the second old seating environment, which comprised a seat from which the operator controlled the machine using control sticks fixed on a rigid structure beside the operator and connected rigidly to the rigid platform of the shaker table; this seat is called a floor-mounted control. The new seat, with all controls attached to its armrest, represents the new environment; this seat is called seat-mounted control (SM) and is shown in Fig. 4.6C. One accelerometer was rigidly attached to a halo on the subject's head and another was attached to a pad on the interface between the seat cushion and the operator's buttocks; these were used to calculate the transfer functions between the input acceleration and the subject's head. Camera-based passive markers were also attached on the subject body and on the seat and surrounding equipment to measure the displacement during vibration (Fig. 4.6D).

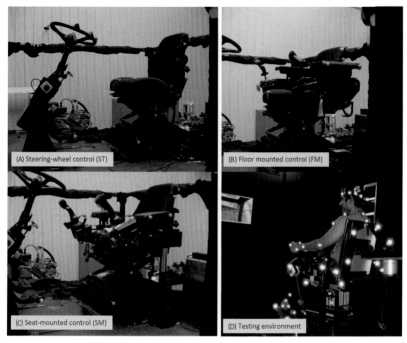

FIGURE 4.6 Testing configurations including (A) steering wheel control (ST), (B) floor-mounted control (FM), (C) seat-mounted control (SM), and (D) a subject sitting in the standard posture and using the armrest control to perform a drilling task on a simulator during whole-body vibration. The *white* dots represent reflective markers used by the motion capture system.

4.2.4.3.1 Methods and data collection

Fig. 4.7 shows the conditions when a subject was using the old seat configuration with the ST (Fig. 4.7A) and another subject was using the new seat configuration with the controls attached on the armrest of the seat SM (Fig. 4.7B). The subjects took different sitting postures similar to their posture in real life: one sat in a standard posture supported by the seat back (SM), and the other sat in a forward, upright, unsupported-back posture (ST). In both cases, the subjects sat with their feet on a foot pedal. A marker-based motion capture system was used to capture the motion of the different segments of the human body as shown in Fig. 4.7A and B.

Five human subjects were used in this study and were tested under different types of vibration rides. The tests consisted of vertical single-axis and three translational (X, Y, and Z) multiple-axis WBV using ride files of 60 seconds in length. The files were recorded from a heavy construction machine, the Caterpillar D10 dozer. A six-degree-of-freedom Servotest (Sears Seating Facility, Davenport, IA, United States) hydraulic motion platform was used to generate the vibration rides.

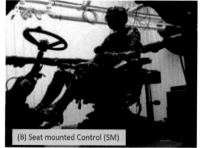

(A) Steering-mounted control (ST)

(B) Seat mounted Control (SM)

FIGURE 4.7 Seat configurations: (A) operator using steering wheel control (ST) and (B) operator using seat-mounted control (SM). *Adapted from Rahmatalla, S., & DeShaw, J. (2011a). Effective seat-to-head transmissibility in whole-body vibration: Effects of posture and arm position.* Journal of Sound and Vibration, 330, 6277−6286.

4.2.4.3.2 Data analysis and results

The transmissibility matrices between the seat and head were calculated for this multiaxis WBV. In this case, the resulting seat-to-head transmissibility comprised many graphs. To compare different seat configurations, graphs in multiple directions were combined using the effective transmissibility concept (Rahmatalla & DeShaw, 2011a). The mean effective transmissibility of the five subjects for the ST and SM configurations during the single-input (SIP) vertical-random and three-output (3OP) translational vibration (SIP−3OP) showed similar characteristics with peaks around 4 Hz (Fig. 4.8). However, it is clear from the figure that with the SM configuration, the subjects showed more head motion than in the ST condition for the frequency range up to 4.5 Hz. The larger head motion of the operators with the SM configuration may trigger an additional sense of discomfort, and that feeling may worsen with time. This result was consistent with the reported discomfort of the subjects. The motion capture data also showed another piece of this puzzle, where the relative motion between the head and torso was larger in the SM than in the ST. Finally, the subjects' torsos experienced less motion with the SM because they used the backrest to support their backs, which meant that most of the energy was going to the subjects' heads. Therefore it is likely that the vibration energy entering the human body via the SM went mostly to the head, generating this extra head motion. Although the ST showed less head motion, it may still not be the best option because the operators were using their muscles to control the machine and were stiffening their necks to reduce head motion. This may trigger fatigue during long-duration WBV exposure. Yet the study revealed that the operators still preferred the old seating environments.

This case study demonstrated that discomfort is a very complicated metric to be quantified in WBV; however, it showed a clear relationship between

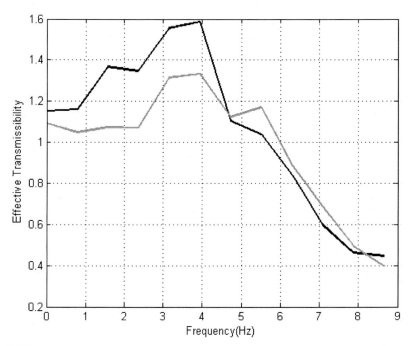

FIGURE 4.8 The mean effective transmissibility between the seat interface and subjects' heads; the dark line represents the seat-mounted control (SM) condition and the *gray* line represents the steering wheel (ST) condition during the SIP–3OP vertical-random vibration.

discomfort and the voluntary and involuntary movements of human body segments during WBV.

4.3 The role of the rotational and translational motions on dynamic discomfort

The previous case study illustrated the importance of including the motion of human body segments in the discomfort evaluation during WBV. As indicated in Chapter 2, Measurement of Human Response to Vibration, measurements of individual human body segments can be achieved with state-of-the-art motion capture systems. It also showed that each segment of the human body can have six degrees of movement: three translational and three rotational. Now the question is this: if discomfort can be affected by the motion of the body segments, can it also be affected by the type and direction of motion? The answer is yes, and that will become clear in the following sections and case study examples.

It is known that the human body, like any other biomechanical system, can be affected by the magnitude and frequency of vibration, among other things. The effect of frequency content can be very important when the vibration generates frequencies that match the natural frequency of the human body (called resonance). Under such frequencies, the human body will generate large motion, bigger than the motion input to the body, which can cause more discomfort and may lead to injury under severe conditions. For example, in their investigations about the resonance frequencies of the human body under WBV, Paddan and Griffin (1994) and Fard et al. (2003) showed that different body segments can resonate at different frequencies in the translational and rotational directions. Hence, different body segments can be affected differently by vibration depending on the natural frequencies of these segments. Therefore it is recommended to measure the motion and relative motion between the human body segments when quantifying discomfort in WBV studies.

The focus of most existing vibration analysis studies on discomfort has been on the effect of translational vibration on the human body segments; the majority paid less attention to the role of rotational motions of the human body. The goal of this section is to show how discomfort relates to the rotational motions of body segments, as well as how this compares to methods that use translational motion components in their discomfort evaluation. Moreover, techniques to calculate and transform the angular motions, similar to those shown for translational motions in Chapter 2, Measurement of Human Response to Vibration, are also presented in this section.

As shown in Chapter 3, Biodynamics of Supine Humans Subjected to Vibration and Shocks, marker-based motion capture systems and inertial sensors are effective tools for measuring the translational and rotational motions of the human body. Most inertial systems on the market provide users with software to assist with calculating the rotational matrices that can be used to transform the motion between the local and global systems and between different local systems on the body, as follows:

$$\begin{bmatrix} \omega_{X_G} \\ \omega_{Y_G} \\ \omega_{Z_G} \end{bmatrix} = \mathbf{R} \begin{bmatrix} \omega_{X_L} \\ \omega_{Y_L} \\ \omega_{Z_L} \end{bmatrix} \quad (4.11)$$

where ω_{X_G}, ω_{Y_G}, and ω_{Z_G} are the angular velocities in the global lab system; ω_{X_L}, ω_{Y_L}, and ω_{Z_L} are the angular velocities in the local coordinate system of the inertial sensors; and $\mathbf{R}$ is the transformation matrix between the local and global systems.

Once the angular velocities for different sensors are aligned with the global coordinate system, the process of finding the relative motions between the local systems becomes straightforward. For example, the relative angular motions between the head and the trunk can be evaluated from the sensors installed at the head and chest as follows:

$$\omega_{X_{Head-Chest}} = \omega_{X_{Head}} - \omega_{X_{Chest}} \quad (4.12)$$

The RMS resultant of the relative angular velocities of each two adjacent segments (i and j) in all three directions (X, Y, and Z), can be evaluated as follows (DeShaw & Rahmatalla, 2016).

$$RMS[\omega_{XYZ}] = \sqrt{RMS(\omega_{X_j} - \omega_{X_i})^2 + RMS(\omega_{Y_j} - \omega_{Y_i})^2 + RMS(\omega_{Z_j} - \omega_{Z_i})^2}$$

$$(4.13)$$

4.3.1   Case study: role of rotational motion in discomfort

This case study (DeShaw & Rahmatalla, 2016) illustrates the role of the rotational motions of the human segments in the resulting perceived discomfort. Twelve male subjects with no history of neck, shoulder, or head injuries, or any neurological conditions, participated in this study. The study involved exposing the subjects to different types of WBV while they were immobilized in supine positions. Fig. 4.2 shows the different immobilization conditions. The WBV was generated by a man-rated shaker that can simulate different ride scenarios. The subjects were exposed to WBV in the three translational directions (vertical in the X-direction, fore-aft in the Z-direction, and side-to-side in the Y-direction), including 3IP with high and low magnitude (3D high and 3D low) and 3IP with high and low magnitude mixed with shocks (3D + S high and 3D + S low). The ride files contained random vibration files with a frequency content of 0.5−25 Hz. Each ride file was run for 60 seconds. The motion was measured by inertial sensors that were attached to the interface between the human body and the supporting surfaces and to the human body on the forehead just between the eyebrows, the sternum, the left anterior superior iliac spine, and the patella of the left knee.

4.3.1.1   Reported discomfort

The subjects reported their discomfort based on the Borg CR-10 scale (Fig. 4.1). They were asked to consider 0 as representing a condition that is equivalent to sleeping in bed and used that as a reference for comparison with other testing conditions. The relationships between the reported discomfort and the lumped percentages of reported discomfort at the head-to-chest and chest-to-pelvis areas are depicted in Fig. 4.4.

4.3.1.2   Relative RMS angular velocity

The results of the relative RMS magnitude of the angular velocity between the adjacent segments of the human body for each condition, calculated based on Eq. (4.13), are demonstrated in Fig. 4.9. These motions include roll, pitch, and yaw for the head-to-chest and chest-to-pelvis. Each bar comprises six slides representing the RMS angular velocity of the head-to-chest roll angle at the bottom of the bar, followed by the head-to-chest pitch angle,

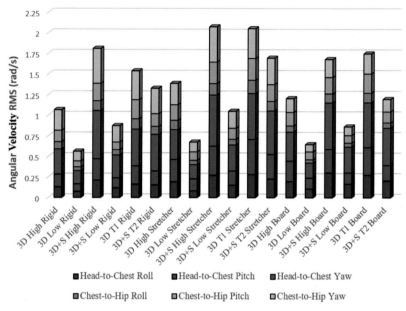

FIGURE 4.9 Average root mean square (RMS) relative angular velocity between human body segments under different immobilization conditions when subjected to different forms of vibration: 3IP with high and low magnitude (3D high and 3D low), 3IP with high and low magnitude mixed with shocks (3D + S high and 3D + S low). Files with the letter T such as 3IP (3D T1 and 3D + S T2) are used during validation. *Data is from DeShaw, J., & Rahmatalla, S. (2016). Predictive discomfort of supine humans in whole-body vibration and shock environments. Ergonomics, 59(4), 568–581. https://doi.org/10.1080/00140139.2015.1083125.*

the head-to-chest yaw angle, the chest-to-hip roll angle, the chest-to-hip pitch angle, and the chest-to-hip pitch angle on the top of the bar. Fig. 4.9 shows that the largest relative angular velocities did not occur with the rigid surface that showed the highest static discomfort (Fig. 4.3), but rather on the stretcher with the 3D vibration with the shock condition (3D + S High Stretcher) that showed the highest angular velocity. As shown in Fig. 4.9, there is a good correlation between the magnitude of the RMS values of the relative angular velocity and the corresponding conditions of the reported discomfort in Fig. 4.3. This indicates that dynamic discomfort has good correlations with the relative angular velocity between the adjacent segments of the human body.

4.4 Predictive discomfort of supine humans

4.4.1 Case study

This case study presents methodologies that use rotational motions of the human body segments, including angular acceleration and angular velocity, as metrics for predicting discomfort in WBV (DeShaw & Rahmatalla, 2014). It also demonstrates the difference between the efficacy of this approach and the methods presented in ISO 2631-1.

As demonstrated in the previous section and case study, the RMS of the relative angular velocity of the adjacent segments of the human body demonstrates good correlations with the reported discomfort. Based on this idea, DeShaw and Rahmatalla (2016, 2011) developed predictive discomfort measures for seated and supine positions in WBV. The proposed discomfort comprises two components of discomfort: static and dynamic. The static discomfort portion is constructed using the reported discomfort of the subjects before vibration according to the Borg CR-10 scale. The dynamic discomfort is a function of the range of motion of the joints and can be affected by several factors, including joint anatomy, joint limits, and severity of motion. Based on the anatomy, the joint can move with a certain range of motion that is considered comfortable; however, discomfort starts to increase exponentially when the joint reaches its active limits, and pain can occur when the angle progresses beyond this point. The reason for considering joint limits is the possibility that human joint motion can be severe when humans are exposed to severe vibrations that contain shocks. This may become important if the formula is applied to injured patients whose joint motions can trigger discomfort and pain during WBV.

The following equation demonstrates one form of the predictive discomfort function (DeShaw & Rahmatalla, 2011; Rahmatalla & DeShaw, 2011b):

$$F(q) = \sum_{i=1}^{j} W_i(\Delta q_i^{norm} + G \times QU_i + G \times QL_i) + \sum_{i=1}^{j} \alpha_i(\Delta \ddot{q}_i) \qquad (4.14)$$

where $F(q)$ is the discomfort as a function of the joint angle (q) and j is the number of joints. The right side of the equation ($W_i(\Delta q_i^{norm} + G \times QU_i + G \times QL_i)$) represents the contribution from quasistatic discomfort, which is discomfort that can be triggered during slow motions such as changing postures. In this form, Δq^{norm} represents the deviation of the joint angle from its neutral posture, when a person is standing and relaxing, and is normalized to be between 0 and 1; $G \times QU$ is a penalty term when the joints approach their upper active limits; $G \times QL$ is a penalty term when the joints approach their lower active limits (DeShaw & Rahmatalla, 2014; Marler, et al., 2009); and W_i represents the weighting parameters related to the contribution of the quasistatic discomfort from each joint. The right side of the equation $\alpha_i(\Delta \ddot{q}_i)$ represents the contribution from the dynamic discomfort, where ($\Delta \ddot{q}_i$) is the relative acceleration between the adjacent segments and α_i are weighting parameters related to the contribution of the dynamic discomfort from each joint. As shown in Eq. (4.14), this form does not include static discomfort but mainly focuses on the relative angular motions of the joints. Eq. (4.15) presents another form that is based on the angular velocity of the joints instead of their angular acceleration. This predictive discomfort function was also proposed by DeShaw and Rahmatalla (2014) and is given as follows:

$$F(q) = \sum_{i=1}^{j} W_i(\Delta q_i^{norm} + G \times QU_i + G \times QL_i) + \sum_{i=1}^{j} \alpha_i(\Delta \dot{q}_i) \quad (4.15)$$

This equation is very similar to Eq. (4.14), except for the difference in the motion components, where $(\Delta \dot{q}_i)$ is used instead of $(\Delta \ddot{q}_i)$. In order to extend the benefit of Eqs. (4.14) and (4.15) and make them comparable to the reported discomfort, the objective functions are transformed to a scale between 0 and 10, similar to those used in the Borg CR-10 scale (DeShaw & Rahmatalla, 2016). The new scaled objective discomfort function takes the following form:

$$F(q)^{Borg} = \sum_{i=1}^{j} W_i(\Delta q_i^{Borg}) + \sum_{i=1}^{j} \alpha_i(\Delta \dot{q}_i^{Borg}) \quad (4.16)$$

The modified form in Eq. (4.16) does not include the joint limit portion shown in Eqs. (4.14) and (4.15). The reason behind discarding the joint limit term is that subjects will rarely reach their joint limits during WBV, especially with supine positions where most joints take their neutral position; still, this term can be brought back to the equation if there is a tendency for this type of motion to take place or has significant effects. The term Δq_i^{Borg} is equivalent to the Borg CR-10 score coming from each joint due to the quasistatic discomfort that is related to postural changes, and $\Delta \dot{q}_i^{Borg}$ is equivalent to the Borg CR-10 scale of the angular velocity of each joint $(\Delta \dot{q}_i)$.

The discomfort function in Eq. (4.16) includes the quasistatic discomfort that may be encountered when slow, voluntary motions such as postural changes take place. It does not include static discomfort, which reflects the discomfort level of the subjects when they are statically sitting or lying on a surface without being subjected to motion or generating any type of motion. Therefore to extend Eq. (4.16) and make it comparable to the reported discomfort during WBV, an additional static term is added to the discomfort function, and the final discomfort function takes the following form:

$$F(q)^{Borg} = F_{Statics}^{Borg} + \sum_{i=1}^{j} W_i(\Delta q_i^{Borg}) + \sum_{i=1}^{j} \alpha_i(\Delta \dot{q}_i^{Borg}) \quad (4.17)$$

On the right side of the equation, the first term represents static discomfort, the second term represents quasistatic or postural discomfort, and the last term represents dynamic discomfort.

4.4.2 Case study: predictive discomfort for supine postures

This case study demonstrates the application of the predictive discomfort function (Eq. (4.17)) on supine human subjects exposed to WBV (DeShaw and Rahmatalla, 2016). The subjects were immobilized using three conditions as shown in Fig. 4.2. First, the static discomfort component was evaluated using the Borg CR-10 scale. This was done by asking the subjects about their discomfort level while they were lying on the different support conditions (rigid, stretcher, and stretcher and spine-board as shown in Fig. 4.2) and before any motion took place. In general, the static discomfort comprises the physical discomfort that can result from the stiffness of the contact surfaces and the pressure points where the human body contacts the supporting surfaces (Ebe & Griffin, 2000a; Kolich, 2004). The subjects were asked to rate their discomfort with 0 representing the value that is equivalent to their discomfort when they are lying on a bed.

Subjects in supine positions are normally immobilized and prevented from voluntary motions, especially during medical transport, so the second term in the discomfort function in Eq. (4.17), which represents the quasistatic discomfort component, was ignored. Accordingly, Eq. (4.17) will be simplified to:

$$F(q)^{Borg} = F_{Statics}^{Borg} + \sum_{i=1}^{j} \alpha_i (\Delta \dot{q}_i^{Borg}) \qquad (4.18)$$

Again, the quasistatic term in Eq. (4.17) should be included in the equation if the subjects occasionally change their postures or make voluntary motions. While it can be complicated to measure the relative motions of each body segment, an initial investigation was conducted to select the joints that have major effects on discomfort and to neglect the values at other joints that contribute less to discomfort. This was done by averaging the reported discomfort from all subjects at most joints of the body. The investigation showed that discomfort was mostly localized at the head-neck region, which contributed to up to 70% of the reported localized discomfort, and the rest of the reported discomfort was mostly in the low back area with a contribution of up to 30% of the reported localized discomfort. Fig. 4.4 shows the bar graphs of the lumped discomfort components. Based on the ratios presented in Fig. 4.4, the values for the weighting factors (α_i) were assigned as 0.7 for the head−neck area and 0.3 for the back area. The resulting predictive discomfort function was simplified further to include the angles between the head and chest (the neck area) and between the chest and pelvis (the lower back area), as follows:

$$F(q)^{Borg} = F_{Statics}^{Borg} + 0.7(\Delta \dot{q}_{Head-Chest}^{Borg}) + 0.3(\Delta \dot{q}_{Chest-Pelvis}^{Borg}) \qquad (4.19)$$

where ($\Delta \dot{q}_{Head-Chest}^{Borg}$) and ($\Delta \dot{q}_{Chest-Pelvis}^{Borg}$) represent the dynamic motion of the joints.

The angular velocity terms in the Borg scale, $\Delta \dot{q}^{Borg}$, are not normally provided from the measurements but should be calculated. One way to circumvent this problem is to determine a relationship between the measured

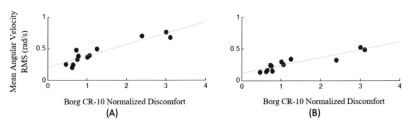

FIGURE 4.10 Relationship between the mean reported discomfort based on the Borg CR-10 scale and the mean relative root mean square (RMS) angular velocity when subjects were exposed to: (A) head-to-chest and (B) chest-to-pelvis. The circular points represent the average of all subjects for the 12 testing conditions. *Data is from DeShaw, J., & Rahmatalla, S. (2016). Predictive discomfort of supine humans in whole-body vibration and shock environments. Ergonomics, 59(4), 568–581. https://doi.org/10.1080/00140139.2015.1083125.*

angular velocity $\Delta\dot{q}$ and the equivalent Borg score $\Delta\dot{q}^{Borg}$ at different joints. Fig. 4.10 shows the correlations between the RMS of the relative angular velocity and the reported discomfort in the Borg CR-10 scale of the 12 subjects during WBV at the head−chest and chest−pelvis areas. The calculations resulted in strong correlation coefficients of $R^2 = 0.93$ for the head-chest region and $R^2 = 0.89$ for the chest−pelvic region.

The slopes and intersections of the regression lines in Fig. 4.10A and B can be used to establish relationships between the relative RMS angular velocity $\Delta\dot{q}$ and the reported discomfort in terms of the Borg CR-10 scale $\Delta\dot{q}^{Borg}$ as follows:

For the head−chest region, the following relationship can be established:

$$\Delta\dot{q}_{Head-Chest} = 0.82\Delta\dot{q}^{Borg}_{Head-Chest} + 0.20 \qquad (4.20)$$

Writing Eq. (4.20) in terms of the Borg CR-10 scale:

$$\Delta\dot{q}^{Borg}_{Head-Chest} = 5.48\Delta\dot{q}_{Head-Chest} - 1.10 \qquad (4.21)$$

For the chest−pelvis region:

$$\Delta\dot{q}_{Chest-to-Hip} = 0.124(\Delta\dot{q}^{Borg}_{Chest-Pelvis}) + 0.114 \qquad (4.22)$$

Writing Eq. (4.22) in terms of the Borg CR-10 scale:

$$\Delta\dot{q}^{Borg}_{Chest-to-Hip} = 8.08(\Delta\dot{q}^{Borg}_{Chest-Pelvis}) - 0.92 \qquad (4.23)$$

The relationships in Eqs. (4.21) and (4.23) are substituted for the discomfort in Eq. (4.19) to present a general discomfort function for supine humans that is expressed in terms of the Borg CR-10 scale (DeShaw and Rahmatalla, 2016).

$$F(q)^{Borg} = F^{Borg}_{Statics} + 0.7(5.48\Delta\dot{q}_{Head-Chest} - 1.10) + 0.3(8.08\Delta\dot{q}_{Chest-Pelvis} - 0.92) \qquad (4.24)$$

Because Eq. (4.24) is derived based on many immobilization and vibration conditions, it is expected that the equation will hold for other conditions, which are not part of the data that were used in the construction of the function. Therefore Eq. (4.24) could be considered a general form for the evaluation of discomfort for supine humans under different vibration conditions. To test this hypothesis, the developed predictive discomfort function (Eq. (4.24)) was tested using data from subjects under different vibration rides but with constraint conditions similar to those used in the initial testing (Fig. 4.2). To this end, 6 of the 12 subjects used in the initial testing were asked to come back to be tested with a new set of vibration conditions, which included 3IP vibration and shocks using rigid support, 3IP vibration using a stretcher (Fig. 4.2B), 3IP vibration and shocks using a stretcher, and 3IP vibration using a spine board (Fig. 4.2C). In Fig. 4.11, the filled circles represent the new testing conditions, while the open circles represent the original testing conditions. The figure shows that the results of the new conditions were very close to the regression line of all initial tests and that they fell within the 95th percentile confidence bands.

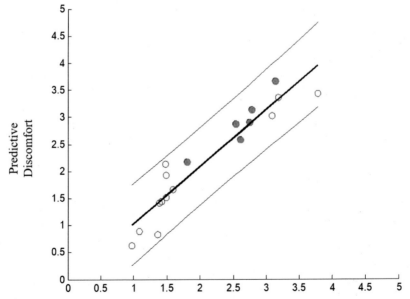

FIGURE 4.11   The correlation and the best line fit (bold line and 95% confidence lines) between the predictive discomfort function and the average reported Borg CR-10 score of 12 subjects. The open circles represent the data from the development phase, and the filled circles represent the average reported Borg CR-10 score of six subjects during the second validation phase. *Data is from DeShaw, J., & Rahmatalla, S. (2016). Predictive discomfort of supine humans in whole-body vibration and shock environments.* Ergonomics, 59(4), 568–581. https://doi.org/10.1080/00140139.2015.1083125.

Fig. 4.11 shows a strong linear trend, with an R^2 value equal to 0.94, a line slope near 1, and an intercept near 0. The all-subjects averaged data gives much better results than the individual subjects, which indicates that the predictive discomfort function may be very effective in predicting the average discomfort of a group rather than of a single person.

4.5 Predictive discomfort of nonneutral postures in seated positions

There are many occupations where people are required to use awkward or nonneutral postures to perform their tasks while they are in a seated position and subjected to WBV. One example is when operators drive a forklift and twist their necks to one side or the back. Another example is in prehospital transport, when medics are forced to take nonneutral postures while monitoring the health of patients (Kroening et al., 2019). The effect of awkward postures in the cabs of grab-unloaders machines for bulk material handling was investigated by Courtney and Chan (1999). The study revealed that operators adopted poor postures because they needed to tilt their heads down and adjust their bodies to see and control their tasks. The study determined that the operators spent 50% of the work cycle time looking down vertically. Such awkward postures led to musculoskeletal discomfort at different locations on their bodies, with 88% reporting issues at the lower back, 81% at the neck, and 50% at the mid-back and shoulders.

Vibration with awkward postures can lead to discomfort and pain. Kittusamy and Buchholz (2004) conducted a literature review about the stress that can be generated by the awkward postures of construction equipment operators in WBV. In a study on helicopter pilots (De Oliviera & Nadal, 2005), researchers found that posture and cyclic vibration forces can lead to discomfort and back pain. Studies on all-terrain vehicle (ATV) drivers found that they frequently use nonneutral rotational positions (Rehn et al., 2005). In their research on professional ATV drivers, Rehn et al. (2005) observed that drivers frequently twist their heads for a short period of time while they were driving on irregular roads that generate WBV. They demonstrated this nonneutral rotational posture of the head−neck region as an ergonomic risk factor. Furthermore, Eger et al. (2008) investigated a similar situation in load-haul-dump operators and found that they drove their machines with their necks rotated more than 40 degrees for over 89% of the operation time. The study found that exposure to vibration with such nonneutral postures can generate risk to injury and that risk is considered beyond the Swedish National Work Injury Insurance Criteria for neck rotation. Finally, in their study in the mining industry, Eger et al. (2008) found that the risk of musculoskeletal injuries could potentially be increased when operators combined the neck rotation with awkward torso postures such as twisted and lateral postures.

4.5.1   Case study: effect of nonneutral postures on discomfort

The effect of WBV on the discomfort of seated people taking nonneutral postures in WBV will be introduced in this case study (DeShaw & Rahmatalla, 2014; Rahmatalla & DeShaw, 2011b). In this case, the predictive discomfort function (Eq. (4.17)) was modified and applied to a seated condition in situations where people use awkward or nonneutral postures while conducting tasks under WBV. The study also compared the results of the developed predictive discomfort function and those predicted by the current standard ISO 2631-1.

4.5.1.1   Study design and data collection

Twelve healthy male subjects participated in this study. The subjects sat on a rigid seat and took different seated postures as shown in Fig. 4.12. Also, the subjects were asked to hold the posture for 60 seconds during each vibration ride without making any voluntary movements. The subjects were exposed to vibration with different magnitudes and directions, including (1) SIP random fore-aft, lateral, and vertical vibrations with an RMS vibration magnitude of 1.8 m/s^2 and frequency content from 0.5 to 12 Hz; (2) 3IP random vibration with an RMS value of 1.1 m/s^2 in each direction; and (3) 6IP with an RMS value of 1.1 m/s^2 in each translational direction and angular accelerations with an RMS value of 0.8 rad/s^2 in the roll, pitch, and yaw directions. Each vibration file lasted 60 seconds.

During the testing and at the end of the ride files, the subjects were asked to rate their discomfort level based on the Borg CR-10 scale, which ranges from 0 to 10, with 0 indicating no discomfort and 10 indicating the most severe discomfort. The Borg CR-10 scale is an absolute scale because it allows for the comparison between multiple postures and vibrational conditions. The subjects were allowed at least 15 seconds to perceive the vibration before rating their discomfort level. At the end of each test, each participant rated the static discomfort due to the rotation at the head during a period of no vibration. In addition to collecting the reported discomfort of the subjects, the motion response of the subjects was also recorded using inertial sensors attached to their heads, the back of the neck at the C7 vertebra, and the pelvis. Another inertial sensor was attached to the rigid seat to measure the input vibration entering the subjects' bodies.

4.5.1.1.1   Data analysis and discomfort function formulation

Eq. (4.17) is shown again here for convenience:

$$F(q)^{Borg} = F_{Statics}^{Borg} + \sum_{i=1}^{j} W_i(\Delta q_i^{Borg}) + \sum_{i=1}^{j} \alpha_i(\Delta \dot{q}_i^{Borg})$$

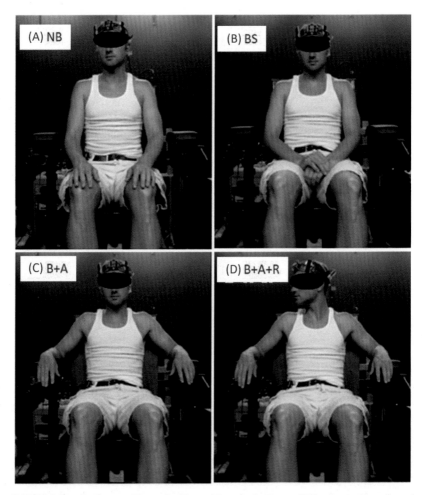

FIGURE 4.12 Seating conditions of subjects: (A) no back support (NB), where subjects leaned forward without touching the seatback, (B) back support (BS), where subjects leaned back and supported their backs with the seatback, (C) back support and armrest (B + A), where subjects supported their backs with the seatback and put their arms on the armrests, and (D) back support and armrest and head rotation (B + A + R), which is similar to (C) but with the head fully twisted to one side but not initiating joint limit discomfort.

The equation is modified here to account for the contribution of awkward postures to the discomfort of seated humans. In this case, the data reported from each subject regarding the static discomfort $F_{Statics}^{Borg}$ were merged with the term that includes the quasistatic discomfort $\sum_{i=1}^{j} W_i(\Delta q_i^{Borg})$ due to the rotation (posture) at the head during a period of no vibration. This data included discomfort coming from the awkward postures and the static discomfort coming from the seat. Based on the initial calculations, the average weight multiplier for the head−chest region was selected to be 0.5. For convenience, the same weighting factors were applied to the dynamic discomfort part, that is, a 0.5 weighting factor was used for both the head−chest and chest−pelvic regions. Based on these adjustments, the predictive objective discomfort function for seated subjects with nonneutral posture takes the following form:

$$F(q)^{Borg} = 0.5(F^{Borg}_{Statics+QuasiStatics} + \Delta\dot{q}^{Borg}_{Head-Chest} + \Delta\dot{q}^{Borg}_{Chest-to-Hip}) \quad (4.25)$$

In an approach similar to that for the supine position, the $\Delta\dot{q}^{Borg}_i$ terms in Eq. (4.25) are transformed to the Borg CR-10 scale. In this case, the RMS angular velocities at the head−chest and the chest−pelvis are plotted with the average reported discomfort values of all vibration conditions including SIP with different vibration magnitudes, three-directional input with low (3IP-L) and high (3IP-H) vibration magnitudes, and six-directional inputs with low (6IP-L) and high (6IP-H) vibration magnitudes.

Fig. 4.13 depicts the relationship between the average Borg CR-10 scores and the RMS angular velocity at the head−chest and chest−pelvis during the SIP and 3IP vibration. Each solid point in the graphs represents one posture during one vibration condition. The resulting least-squares regression lines are shown to have the following forms.

For the head−chest region:

$$\Delta\dot{q}_{Head} = 0.14\Delta\dot{q}^{Borg}_{Head} - 0.14 \quad (4.26)$$

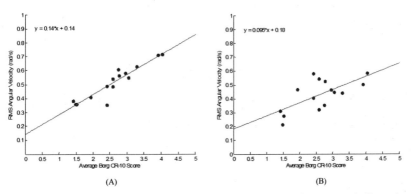

(A)   (B)

FIGURE 4.13 The relationship between the root mean square (RMS) average of all angular velocity components and the average Borg CR-10 scores of all subjects when exposed to SIP and 3IP vibration for (A) the head−chest region and (B) the chest−pelvis region. Each of the solid circles represents one posture during one vibration condition. The regression lines are $y = 0.14x−0.14$ for the head−chest region with ($R^2 = 0.931$) and $y = 0.095x$ 0.18 for the chest−pelvis region with ($R^2 = 0.672$). *SIP*, Single input; *3IP*, three-directional input. *Adapted from DeShaw, J., & Rahmatalla, S. (2014). Predictive discomfort in single- and combined-axis whole-body vibration considering different seated postures. Human Factors, 56(5), 850−863.*

Transforming Eq. (4.26) to the Borg scale:

$$\Delta \dot{q}_{Head}^{Borg} = 7.1428^* \Delta \dot{q}_{Head} - 1.0 \qquad (4.27)$$

For the chest—pelvis region:

$$\Delta \dot{q}_{Spine} = 0.095 \Delta \dot{q}_{Spine}^{Borg} + 0.18 \qquad (4.28)$$

Transforming Eq. (4.26) to the Borg scale:

$$\Delta \dot{q}_{Spine}^{Borg} = 10.5263^* \Delta \dot{q}_{Spine} - 1.895 \qquad (4.29)$$

Using the previous relationships (Eqs. (4.27) and (4.29)), the final form of the predictive discomfort equation for the seated humans with nonneutral postures in terms of Borg CR-10 units is as follows (DeShaw and Rahmatalla, 2014):

$$f^{Borg}(q) = 0.5(F_{Statics+QuasiStatics}^{Borg} + (7.1428 \times \Delta \dot{q}_{Neck} - 1.0) + (10.5263 \times \Delta \dot{q}_{Lower-back} - 1.895))$$
$$(4.30)$$

While it is expected that the angular velocity at the neck or the lower-back regions may be exposed to more than one direction, the resultant angular velocity in three rotational directions is used in the evaluation of discomfort under such cases.

4.5.1.1.2 Results

Fig. 4.14 shows the predictive discomfort of the 12 subjects for each posture during the single-axis (SIP) and 3D (3IP) vibration exposures.

Because the predictive discomfort function was transformed with all equivalent Borg CR-10 units, both axes in Fig. 4.14 have a Borg CR-10 scale. A high correlation is seen between all test conditions, with $R^2 = 0.9273$. A slope near 1.0 indicates high correlations between the predicted discomfort and the reported discomfort rating of the participants. The round points in Fig. 4.14 represent every posture during every SIP vibration condition in the fore-aft, lateral, vertical, (3IP-low) 3D-L, and (3IP-high) 3D-H directions. The diamond points represent every posture during the (6IP-low) 6D-L and the (6IP-high) 6D-H vibration conditions. The coefficients of the regression line passing through these round points are used to determine the factors relating the RMS of the angular velocity and the reported discomfort. When the predictive discomfort function is tested using data from new 6D-L and 6D-H vibration conditions, the new tested diamond points fit in very well with the original data set. The diamond points represent each of the four postures in the 6D-L and 6D-H vibration conditions.

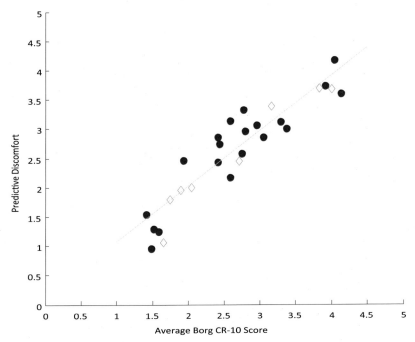

FIGURE 4.14 Average predictive discomfort compared to the Borg CR-10 score of 12 subjects during 28 posture and vibration conditions. The 20 *blue* circular points represent every posture (NB: no back support; BS: back support; B + A: back support and armrest; B + A + R: back support and armrest and head rotation) during each SIP vibration condition in the fore-aft, lateral, vertical, 3IP-low (3D-L) and 3IP-high (3D-H) conditions. The eight diamond-shaped points represent every posture during (6IP-low) 6D-L and (6IP-high) 6D-H vibration conditions. The regression line is also shown with a slope near 1.0 and $R_2 = 0.9273$. *Adapted from DeShaw, J., & Rahmatalla, S. (2014). Predictive discomfort in single- and combined-axis whole-body vibration considering different seated postures.* Human Factors, 56(5), 850−863.

4.5.1.1.3 Predictive discomfort and posture results

Fig. 4.15 shows a bar diagram of the reported and predicted discomfort, both evaluated based on the Borg CR-10 scale of the 12 subjects when they were exposed to 6IP random vibration with high 6D-H and low 6D-L vibration magnitudes. It is obvious from Fig. 4.15 that the values of the predicted and reported discomfort are comparable across all vibration and posture configurations. The highest discomfort magnitudes are seen with the B + A and B + A + R configurations. The higher head motion with the BS as explained in the section above may play a role in this process. The lowest discomfort was associated with the NB, which is expected as demonstrated in an earlier section (Figs. 4.7 and 4.8).

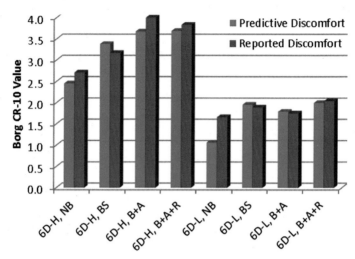

FIGURE 4.15   Comparison between the average reported discomfort and predictive discomfort based on the Borg CR-10 scale for all 12 subjects in different postures (NB: no back support; BS: back support; B + A: back support and armrest; B + A + R: back support and armrest and head rotation) and vibration conditions (6IP-H high magnitude and 6IP-L low magnitude). *Data from DeShaw, J., & Rahmatalla, S. (2014). Predictive discomfort in single- and combined-axis whole-body vibration considering different seated postures.* Human Factors, *56(5), 850–863.*

4.5.2   Predictive discomfort and ISO standard

This section presents a comparison study to show the difference between discomfort values evaluated using the frequency-weighted acceleration values as outlined in the ISO-2631-1 and a methodology similar to that of Marjanen and Mansfield (2010), and the values predicted by the predictive discomfort function Eq. (4.30). The different vibration conditions, including the SIP vibration condition in the fore-aft, lateral, and vertical directions, (3IP-low) 3D-L and (3IP-high) 3D-H, and (6IP-low) 6D-L and (6IP-high) 6D-H vibration conditions, were used in each frequency calculation. Consideration was also taken for the contact points (footrest, seat pan, and backrest) for each of the four postures. It should be noted that the only postural distinction in the ISO-2631-1 is between the NB and BS postures. Therefore the 28 posture and vibration conditions resulted in only 14 unique frequency weighted accelerations values. The other 14 combinations, which included the B + A and B + A + R postures, yielded the same result as the simply supported BS posture. As mentioned before, ISO 2631-1 uses frequency-weighted accelerations to estimate discomfort. Fig. 4.16 shows the comparison between the average reported discomfort based on the Borg CR-10 scale and the discomfort estimated by ISO. The discomfort estimated by ISO, represented by a wide *red* block containing the three (*blue* squares) postures (BS, B + A, and B + A + R), showed a correlation with $R^2 = 0.8951$. However, because

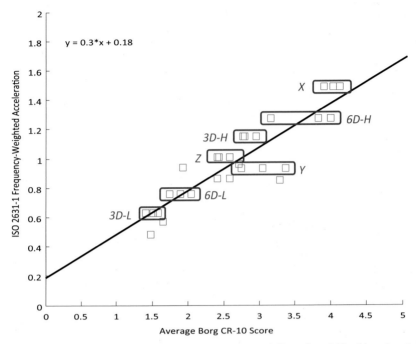

FIGURE 4.16 The relationship between the average reported discomfort of 12 subjects based on the Borg CR-10 scale and estimated discomfort based on ISO 2631-1 frequency-weighted accelerations. The *red* wide rectangle represents the estimated values by ISO for the different vibration (SIP in *X, Y,* and *Z,* 3IP-L, 3IP-H, 6IP-H, and 6IP-L) and posture (BS: back support; B + A: back support and armrest; B + A + R: back support and armrest and head rotation) conditions. *SIP,* Single input; *3IP-L,* three-directional input with low; *3IP-H,* three-directional input with high; *6IP-H,* six-directional input with high; *6IP-L,* six-directional input with low. *Data from DeShaw, J., & Rahmatalla, S. (2014). Predictive discomfort in single- and combined-axis whole-body vibration considering different seated postures.* Human Factors, 56(5), 850−863.

nonneutral postures are not part of the ISO evaluation, the discomfort estimated by ISO did not capture the effect of posture for the different configurations. Although the ISO used acceleration information from the input and did not capture the effect of posture, the results showed that it can have reasonable predictions.

While the correlation is high between the reported discomfort and the ISO 2631-1 prediction, Fig. 4.16 shows that ISO 2631-1 only accounts for the backrest and footrest contacts and not other postures. This can be seen in Fig. 4.16 where the three postures, represented by *blue* squares inside the wide *red* rectangles, resulted in the same frequency-weighted accelerations values predicted by the standard; this means that the three postures (BS, B + A, B + A + R) gave the same frequency-weighted acceleration values. So, in this study, only 14 unique frequency weightings are generated out of the 28 vibration and posture conditions, due to the three postures being repeated by the standard (BS, B + A, B + A + R). For example, in 6D-L vibration, the current standards estimate a frequency-weighted acceleration of around 0.75 m/s^2 for each of the BS, B + A, and B + A + R postures due to the input accelerations. This is because the standard considered the seat and footrest to be the same and disregarded the effect of posture.

4.5.3   Predictive discomfort using the angular acceleration versus angular velocity

The previous case study on the effect of nonneutral postures on discomfort also investigates the efficacy of using angular acceleration versus angular velocity in the predictive discomfort evaluation (DeShaw & Rahmatalla, 2014). To this end, the correlation between the reported discomfort based on the Borg CR-10 scale is plotted with the RMS angular velocity and the RMS angular acceleration at the head-chest and chest-pelvic areas as shown in Fig. 4.17. Fig. 4.17C and D shows that the data from the angular acceleration are more scattered than the data from the angular velocity (Fig. 4.17A and B). Also, the data for the angular acceleration are further from the regression line that represents the line of best fit. Because the relationships presented in Fig. 4.17 are the bases for constructing the predictive discomfort function, it is clear that angular velocity is a better candidate than angular acceleration for the predictive discomfort function.

4.6   Chapter summary

There are many situations in which humans are exposed to multiaxis vibration while taking different postures. Studies have shown that humans can sense a great deal of discomfort and can be prone to harm when using awkward postures during WBV with excessive magnitudes. Many methodologies have been developed and tested during the last century, including subjective reported discomfort and predictive discomfort. Predictive discomfort metrics have received more attention, and some of these methods were documented in the standards, including ISO 2631-1 and BS 6841:1987. The standards give general guidance on how people perceive vibration; they use the magnitude of the input accelerations between the human and the contact surfaces as a means for evaluating discomfort. Because human discomfort appears to be sensitive to the frequency content and direction of vibration, the standards created different weighting factors to adjust the input vibration magnitude to accommodate these factors. Although there are ongoing concerns about the validity of the weighting multipliers in the standards including postures, they have shown reasonable results in many applications.

  In addition to the methods outlined in the standards, many others have been developed to objectively quantify sensation and discomfort. Most of these methods use modified versions of concepts used in quantifying the

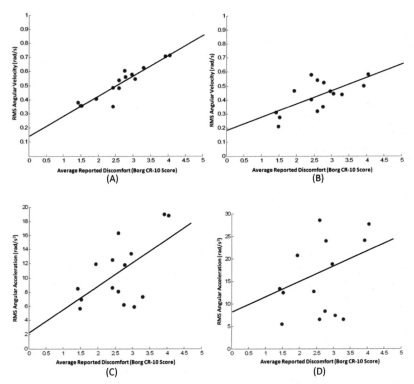

FIGURE 4.17   The correlation between the average reported discomfort based on the Borg CR-10 scale and the root mean square (RMS) of the angular motion at the head–chest and chest–pelvic regions when subjects were exposed to SIP and 3IP vibrations using (A) angular velocity at the head–chest, (B) angular velocity at the chest–pelvis, (C) angular acceleration the head–chest, and (D) angular acceleration at the chest–pelvis. *Adapted from DeShaw, J., & Rahmatalla, S. (2014). Predictive discomfort in single- and combined-axis whole-body vibration considering different seated postures. Human Factors, 56(5), 850–863.*

perception and sensation of sounds and acoustics. The use of the power law to predict discomfort is one popular approach. With the power law, discomfort can be quantified as a function of the vibration magnitude, which is proportional to the power factor of the vibration magnitude (n) and a multiplication factor (k). One limitation of using the power law to quantify discomfort is that the power factor (n) is constant, and therefore the power law will predict the same discomfort value under vibrations with different frequencies. While the latter contradicts the ongoing research that has shown that human discomfort is sensitive to the frequency content of vibration, work has been done to present the power (n) as a function of frequency. With n as a function of the frequency, graphs of contour lines of constant discomfort or sensation have been developed as functions of the vibration magnitude and frequency.

While existing approaches to quantify discomfort are becoming effective in many applications, there are still major challenges to be resolved. For example, most existing methods are unable to describe the actual instantaneous perceptions due to WBV, do not consider static discomfort, and show limited success when it comes to posture and multidirectional vibration. Most importantly, existing methods consider the input motions to the human body as means for evaluating perception and discomfort, and very little work has focused on developing discomfort functions that correlate with the resulting motion of the human body segments during WBV.

Methods that consider the motion of the human body segments as a means to quantify discomfort have shown to be effective when applied to different vibration scenarios, including multiple-axis vibration, different frequencies, different static discomfort, and different postures. These methods have shown very promising results for applications with seated and supine subjects when compared with the reported discomfort. The methods also showed superior results when compared with ISO 2631-1 for cases that involved awkward postures. Transforming these objective methods in terms of equivalent Borg CR-10 units, which is extremely useful for predicting discomfort in real-world applications, has been shown to be possible in any vibration scenario.

References

Arnold, J. J., & Griffin, M. J. (2018). Equivalent comfort contours for fore-and-aft, lateral, and vertical whole-body vibration in the frequency range 1.0 to 10 Hz. *Ergonomics, 61*(11), 1545–1559. Available from https://doi.org/10.1080/00140139.2018.1517900.

Basri, B., & Griffin, M. J. (2013). Predicting discomfort from whole-body vertical vibration when sitting with an inclined backrest. *Applied Ergonomics, 44*, 423–434.

Borg, G. (1982). A category scale with ratio properties for intermodal and interindividual comparisons. In H. G. Geissler, & P. Petzold (Eds.), *Psychophysical judgment and the process of perception* (pp. 25–33). Berlin: Deutscher Verlag der Wissenschaften.

British Standards Institution. (1987). Measurement and evaluation of human exposure to whole-body mechanical vibration and repeated shock *(BS 6841:1987)*. London: British Standards Institution.

Courtney, A. J., & Chan, A. H. S. (1999). Ergonomics of grab unloaders for bulk materials handling. *International Journal of Industrial Ergonomics, 23*, 61–66.

De Oliviera, C. G., & Nadal, J. (2005). Transmissibility of helicopter vibration in the spines of pilots in flight. *Aviation, Space, and Environmental Medicine, 76*, 576–580.

DeShaw, J., & Rahmatalla, S. (2011). Predictive discomfort and seat-to-head transmissibility in low-frequency fore-aft whole-body vibration. *Journal of Low Frequency Noise, Vibration and Active Control, 30*(3), 185–196.

DeShaw, J., & Rahmatalla, S. (2014). Predictive discomfort in single- and combined-axis whole-body vibration considering different seated postures. *Human Factors, 56*(5), 850–863.

DeShaw, J., & Rahmatalla, S. (2016). Predictive discomfort of supine humans in whole-body vibration and shock environments. *Ergonomics, 59*(4), 568–581. Available from https://doi.org/10.1080/00140139.2015.1083125.

Dickey, J. P., Oliver, M. L., Boileau, P. E., Eager, T. R., Trick, L. M., & Edwards, A. M. (2006). Multi-axis sinusoidal whole-body vibrations: Part 1—How long should the vibration and rest exposures be for reliable discomfort measures? *Journal of Low Frequency Noise, Vibration and Active Control, 25*, 175−184.

Ebe, K., & Griffin, M. J. (2000a). Qualitative models of seat discomfort including static and dynamic factors. *Ergonomics, 43*(6), 771−790.

Ebe, K., & Griffin, M. J. (2000b). Quantitative prediction of overall seat discomfort. *Ergonomics, 43*(6), 791−806.

Eger, T., Stevenson, J., Callaghan, J. P., Grenier, S., & VibRG. (2008). Predictions of health risks associated with the operation of load-haul-dump mining vehicles: Part 2—Evaluation of operator driving postures and associated postural loading. *International Journal of Industrial Ergonomics, 38*, 801−815.

Fairley, T. E., & Griffin, M. J. (1988). Predicting the discomfort caused by simultaneous vertical and fore-and-aft whole-body vibration. *Journal of Sound and Vibration, 124*(1), 141−156.

Fard, M. A., Ishihara, T., & Inooka, H. (2003). Dynamics of the head−neck complex in response to the Trunk horizontal vibration: Modeling and identification. *ASME Journal of Biomechanical Engineering, 125*, 533−539.

Goswami, I., Redpath, S., Langlois, R. G., Green, J. R., Lee, K. S., & Whyte, H. E. A. (2020). Whole-body vibration in neonatal transport: A review of current knowledge and future research challenges. *Early Human Development, 146*(2020), 105051.

Griffin, M. J. (1990). *Handbook of human vibration*. London: Academic Press.

Hacaambwa, T. M., & Giacomin, J. (2007). Subjective response to seated fore-and-aft direction whole-body vibration. *International Journal of Industrial Ergonomics, 37*(1), 61−72.

Holmlund, P., Lundström, R., & Lindberg, L. (2000). Mechanical impedance of the human body in the vertical direction. *Applied Ergonomics, 31*, 415−422.

Hwang, C. L., & Yoon, K. (1981). *Multiple attribute decision making: Method and application*. New York: Springer.

International Organization for Standardization. (1997). ISO 2631-1. *Mechanical vibration and shock—Evaluation of human exposure to whole-body vibration—Part 1: General requirements* (ISO Standard No. 2631-1:1997). https://www.iso.org/standard/7612.html.

International Organization for Standardization. (2004). *Mechanical vibration and shock—Evaluation of human exposure to whole-body vibration—Part 5: Method for evaluation of vibration containing multiple shocks* (ISO Standard No. 2631-5: 2004). https://www.iso.org/standard/35595.html.

Johanning, E., Fischer, S., Christ, E., Gores, B., & Landsbergis, P. (2002). Whole-body vibration exposure in United States Railroad Locomotives—An ergonomic risk assessment. *AIHA Journal, 653*(4), 439−446.

Kaneko, C., Hagiwara, T., & Maeda, S. (2005). Evaluation of whole-body vibration by the category judgment method. *Industrial Health, 43*, 221−232.

Kittusamy., & Buchholz, B. (2004). Whole-body vibration and postural stress among operators of construction equipment: A literature review. *Journal of Safety Research, 35*(3), 255−261.

Kolich, M. (2004). Predicting automobile seat comfort using a neural network. *International Journal of Industrial Ergonomics, 33*, 285−293.

Kroening, L., Conti, S., Kinsler, R., Lloyd, A., & Molles, J. (2019). Optimal physical space for en route care: Medic posture and injury survey. In *Poster presentation at Military Health System Research Symposium*, Kissimmee, FL.

Lee, J. H., Jin, B. S., & Yonggu, J. (2009). Development of a structural equation model for ride comfort of the Korean high-speed railway. *International Journal of Industrial Ergonomics, 39*, 7−14.

Lewis, C. H., & Griffin, M. J. (1998). A comparison of evaluations and assessments obtained using alternative standards for predicting the hazards of whole-body vibration and repeated shocks. *Journal of Sound and Vibration Volume, 215*(4), 915−926.

Liu, C., Thompson, D., & Griffin, M. J. (2019). Effect of train speed and track geometry on the ride comfort in high-speed railways based on ISO 2631-1. *Proceedings of the Institution of Mechanical Engineers, Part F: Journal of Rail and Rapid Transit.* Available from https://doi.org/10.1177/0954409719868050.

Mansfield, N. J. (2005). *Human response to vibration.* CRC Press.

Mansfield, N. J., & Maeda, S. (2011). Subjective ratings of whole-body vibration for single- and multi-axis motion. *The Journal of the Acoustical Society of America, 130*(6), 3723−3728.

Mansfield, N. J., Holmlund, P., & Lundstorm, R. (2000). Comparison of subjective response to vibration and shock with standard analysis methods and absorbed power. *Journal of Sound and Vibration, 230*(3), 477−491.

Mansfield, N. J., Mackrill, J., Rimell, A. N., & MacMull, S. J. (2014). *Combined effects of long-term sitting and whole-body vibration on discomfort onset for vehicle occupants.* Hindawi Publishing Corporation International Scholarly Research Notices, Volume 2014, Article ID 852607. Available from http://doi.org/10.1155/2014/852607.

Mansfield, N. J., & Maeda, S. (2007). The apparent mass of the seated human exposed to single-axis and multi-axis whole-body vibration. *Journal of Biomechanics, 40*, 2543−2551.

Mansfield, N. J., & Maeda, S. (2005). Effect of Backrest and Torso Twist on the apparent mass of the seated body exposed to vertical vibration. *Industrial Health, 43*(3), 413−420.

Marjanen, Y., & Mansfield, N. J. (2010). Relative contribution of translational and rotational vibration to discomfort. *Industrial Health, 48*, 519−529.

Marler, R., Arora, J., Yang, J., Kim, H.-J., & Malek, K. (2009). Use of multi-objective optimization for digital human posture prediction. *Engineering Optimization, 41*(10), 925−943.

Miwa, T. (1967). Evaluation methods for vibration effect. I. Measurements of threshold and equal sensation contours of whole-body for vertical and horizontal vibrations. *Industrial Health, 5*, 183−205.

Miwa, T., & Yonekawa, Y. (1969). Evaluation methods for vibration effect. Part 9. Response to sinusoidal vibration at lying posture. *Industrial Health, 7*, 116−126.

Morioka, M., & Griffin, M. J. (2006). Magnitude-dependence of equivalent comfort contours for fore-and-aft, lateral, and vertical hand-transmitted vibration. *Journal of Sound and Vibration, 295*(3−5), 633−648.

Oborne, D. J. (1983). Whole-body vibration and International Standard ISO 2631: A critique. *Human Factors, 25*(1), 55−69.

Paddan, G. S., & Griffin, M. J. (1994). Transmission of roll and pitch seat vibration to the head. *Ergonomics, 37*(9), 1513−1531.

Peng, B., Yang, Y., & Luo, Y. (2009). Modeling and simulation on the vibration comfort of railway sleeper carriages. In *Proceedings of Second International Conference on Transportation Engineering*, pp. 3766−3771.

Rahmatalla, S., & DeShaw, J. (2011a). Effective seat-to-head transmissibility in whole-body vibration: Effects of posture and arm position. *Journal of Sound and Vibration, 330*, 6277−6286.

Rahmatalla, S., & DeShaw, J. (2011b). Predictive discomfort of non-neutral head-neck postures in fore-aft whole-body vibration. *Ergonomics, 54*, 263−272.

Rehn, B., Nilsson, T., Olofsson, B., & Lundstorm, R. (2005). Whole-body vibration exposure and non-neutral neck postures during occupational use of all-terrain vehicles. *The Annals of Occupational Hygiene, 49,* 267−275.

Reiher, H., & Meister, F. J. (1931). The effect of vibration on people (in German). In *Forschung auf dem Gebeite des Ingenieurwesens, 2(α), 381. Translation: Report No. F-TS-616-RE, Headquarters Air Material Command,* Wright Field, OH, 1946.

Stevens, S. S. (1957). On the psychophysical law. *Psychological Review, 64*(3), 153−181.

Zhou, Z., & Griffin, M. J. (2014). Response of the seated human body to whole-body vertical vibration: Discomfort caused by sinusoidal vibration. *Ergonomics, 57*(5), 714−732. Available from https://doi.org/10.1080/00140139.2014.898799.

Chapter 5

Justification and efficacy of prehospital immobilization systems

5.1 Introduction

Millions of trauma patients with potential spine-related injuries or other forms of severe injuries are transported to medical facilities each year by air or ground (Frohna, 1999; Kang & Lehman, 2011). It is estimated that close to 5 million injured patients are subjected to spinal immobilization each year in the United States (Frohna, 1999; Morrissey et al., 2014). During prehospital transport, especially when conducted in off-road conditions, the transport medical vehicle can be exposed to severe translational (up and down, side-to-side, and fore-aft) and rotational (rolling, pitching, and yawing) vibrations and shocks in the six degrees of freedom. These types of motions could generate considerable forces and involuntary motions to the patients and can cause discomfort, pain, bleeding, and secondary injuries.

Field spinal immobilization using a backboard and cervical collar has been standard practice for patients with suspected spine injury for many decades (Podolsky et al., 1983; Young, 1965). Advanced Trauma Life Support courses recommend immediate neck immobilization for trauma patients with suspected spinal injury (Hadley et al., 2001). Normally, these guidelines are established based on expert opinion rather than definite evidence. The current protocols are also based on historical rather than scientific findings (Kang & Lehman, 2011). Spinal immobilization systems are based on a principle that uses mechanical systems to prevent the movement of the body, with more focus on the spinal regions. The goal is to reduce the potential of secondary injury for patients with suspected spinal cord injury. Common immobilization equipment involves the use of a rigid cervical collar, head support blocks, a long board with straps, a litter, a vacuum mattress, a scoop stretcher, a Kendrick Extrication Device, and a pelvic sling, with straps or tape on the head and the rest of the body. Critical reviews of the literature of immobilization systems and techniques (Cuthbertson & Weinstein, 2020; De Lorenzo, 1996) and of the benefits and risks of using prehospital spinal immobilization for trauma patients with suspected spinal cord injury showed

Prehospital Transport and Whole-Body Vibration. DOI: https://doi.org/10.1016/B978-0-323-90103-1.00006-1

185

problems and complications with using immobilization systems and did not find reliable sources to prove the benefits of patient immobilization. Early studies (Chiu et al., 2001), however, showed the importance of proper first aid to avoid secondary injuries due to mishandling patients before reaching the hospital. Therefore more studies related to prehospital transport and immobilization techniques are needed.

The prehospital transport of injured patients can be very complicated and encompasses many scenarios, resulting in a variety of forces in all possible directions. For example, immobilization of a patient in a mountainous or disaster area may require transferring the patient while walking and/or running. In this process, the patient can be subjected to sudden shocks and global body movements due to the irregularity of the pace and the forces applied by the people holding the patient on the transport system. This may be followed by transferring the patient to a medical ambulance. In this case, the patient and the transport system may be subjected to unbalanced lifting that can cause jerky and unbalanced rotational motions. Also, the patient and stretcher could be exposed to impact forces during the process of handling and positioning the immobilization system inside the transport vehicle. A new cycle of whole-body vibration (WBV) will then take place during vehicle transport. Finally, the patient may be taken out of the vehicle and exposed to another handling process during transport to another vehicle and/or the hospital. The point is that there are many uncertainties as to the type of motion/forces to which the patient may be exposed during prehospital transport.

The most direct evidence for potentially harmful effects from movements during patient transport comes from studies of involuntary repetitive movements that occur in vibrational environments, that is, WBV. Both animal and human studies show that WBV can induce pain (Zeeman et al., 2015), exacerbate injury, and cause damage at the cellular level (Perremans et al., 2001; Zeeman et al., 2015). In addition, a study on human subjects during helicopter transport suggested that the level of vibration transmitted to the patient may cause bleeding in unstable bone fractures, such as those involving the pelvis (Carchietti et al., 2013). DeShaw and Rahmatalla (2014) also showed that involuntary repetitive movements under WBV caused discomfort at different locations on the human body, with more intense discomfort at the cervical and lumbar areas. In all of these studies, the level of adverse effects was proportional to the intensity of the motion.

In fact, the current studies showed an ongoing debate between experts on the validity, significance, effectiveness, benefits, and drawbacks of the existing immobilization systems for patients with potential spinal injuries. A major point in this discussion is the lack of concrete evidence about the effectiveness of immobilization systems regarding the medical outcome of the patients, and therefore many researchers are questioning the validity of using these systems. Many of these discussions are based on experience, personal judgment, and experimentations. Unfortunately, most of the experimentations that were used to determine the effectiveness of immobilization systems were conducted in inappropriate environments, mostly static or quasistatic, that do not represent real prehospital transport and that neglect the dynamics of the immobilization systems and the human. Because of the dynamic nature of motion during prehospital transport, immobilization systems behave like any other mechanical system and perform differently than they do in a static environment. For example, most of the work in the literature to evaluate the efficacy of backboards was done under static conditions. By simply placing a person on a backboard, researchers found many adverse effects and inferiority when compared to the performance of a vacuum mattress (Hamilton & Pons, 1996; Lovell & Evans, 1994; Luscombe & Williams, 2003). When tested under dynamics conditions, however, the backboard showed very useful advantages over the vacuum mattress (Mahshidfar et al., 2013; Rahmatalla et al., 2019). This example may explain some of the confusion and contradictory findings in the literature. More discussions about this and other relevant topics will be presented in the coming sections of this chapter.

The development and the state of the art in prehospital transport and immobilization systems comprise many transport systems and products that are widely used in practice. The military, for example, has its own transport and immobilization systems that accommodate their needs and provide ease of use in harsh environments. The civilian sectors also use various types of systems in ground and aerial transport. While the existence of many transport and immobilization systems can provide more options and choices that can serve patients' needs and provide safety during transport, the varying performance of these systems in real-life scenarios can make it difficult for medics and emergency responders to select the right transport and immobilization system for patients with suspected spinal cord injuries. The objective of this chapter is to shed some light on the reasons for the contradictory findings in the literature regarding the effectiveness of existing spinal immobilization systems. The chapter will also demonstrate the performance of some popular transport and immobilization systems that are currently used in the civilian and military sectors with a specific focus on the cervical and lumbar spine areas. The goal here is not to advocate for a certain system, but to show how these systems perform under different vibration ride scenarios that can take place en route to a hospital.

This chapter comprises four sections. In the first, a general discussion sheds some light on the ongoing debate regarding the adequacy of using spinal immobilization systems for trauma patients. The second section discusses the role of immobilization systems in reducing the motion that can occur at different regions of the human body. The third section is devoted to immobilization systems that are applied to the head—neck region of the body and uses study examples to compare the efficacy of some popular immobilization systems. The fourth section discusses the role of immobilization systems in stabilizing the lower back and the performance of some systems that are currently used in the civilian and military fields. The chapter concludes with a summary section.

5.2 Ongoing debate

Let us start this debate with a statement that everyone agrees on: trauma patients should be transported from the point of injury to the closest medical center as quickly and safely as possible. While there are many ways to make it quick, the question is how to make it safe without comprising the patient's health outcomes before reaching the treatment center. People who are against immobilization cannot be convinced of its benefit without seeing proof; they believe that immobilization uses up valuable time and could result in adverse physiological and psychological issues for the patient. People who are proimmobilization mostly follow the traditional guidelines and believe common sense dictates that excessive motions during transport could cause complications and secondary injuries. There is no doubt that everyone working in this area cares deeply about patients' safety and wellbeing, regardless of their opinions about immobilization. The question that should be posed to both groups is the following: If immobilization systems can be developed to be quickly applied and can provide safety and comfort to patients during prehospital transport, will everyone accept them? If the answer is yes, then there is something wrong with the current immobilization systems. One problem is inappropriate design that does not consider human factors and human dynamics as essential parameters in the design process. The second problem is a lack of field and simulated field testing and evaluation of most of these systems using objective measures. If the latter issues are the case, then the problem lies in the design and testing of existing immobilization systems and not in the idea and basic principles behind immobilization.

One goal of this section is to present some contradictory results from the literature and show how these results affect the evolving judgments and perceptions of people working in the area of prehospital transport. In the work of Chan et al. (1996), healthy human subjects were tested under two conditions: a standard backboard and a mattress-splint immobilization. After being exposed to spinal immobilization for 30 minutes, subjects undergoing spinal immobilization with the backboard developed pain at several anatomic regions, while those on the vacuum mattress did not. The conclusion of this study was to recommend the vacuum mattress over the backboard. On the other hand, Mahshidfar et al. (2013) compared the performance of the backboard against the vacuum mattress in trauma victims when they were transported by an Emergency Medical Services system. They found that the backboard offered a significant improvement in comfort over a vacuum mattress for patients with a suspected spinal injury and that immobilization using the backboard was easier, faster, and more comfortable for the patient. They also concluded that the backboard provided an additional decrease in spinal movement as compared to the vacuum mattress. A study by Rahmatalla et al. (2019) using dynamics simulation also showed that the implementation of the backboard was faster than the vacuum mattress and that the backboard performed better than the vacuum mattress in reducing the motions at the cervical spine using dynamics rides that simulated ground and aerial prehospital transport. However, the study showed that the vacuum mattress performed better than the backboard at the lower back area of the body (Rahmatalla et al., 2018).

The contradictory results of the studies cited above are an example of the common disagreement in the literature. It is important to note that, while contradictory results can be confusing, the results presented in these studies are all genuine. A main reason for the conflicting results is how the immobilization systems were tested in the studies. In the first study (Chan et al., 1996), the authors tested the subjects by leaving them lying statically on the backboard and the vacuum mattress. This static test may not represent real situations; when these systems function in dynamic environments, the response and performance can be dramatically different than that found in static testing. In the second and third studies (Mahshidfar et al., 2013; Rahmatalla et al., 2019), the backboard performed better than the vacuum mattress when tested in dynamic environments. This is just one example demonstrating why people have different perceptions about immobilization systems. To avoid this, testing should be done to adequately replicate real-life applications.

Because we are at this point in time where immobilization systems are in use in the field, it is best to make sure that they are working appropriately to serve patients and not cause any harm. One way to achieve this is to follow the excellent work that has been done in the literature and determine the issues with current immobilization systems that need to be solved. While there are medical guidelines for the recommendation and use of immobilization systems, there should also be technical guidelines to regulate their use. In a recent systematic review of the literature, Maschmann and others from Denmark (Maschmann et al., 2019) graded the strength of the evidence, clinical judgment, and a consensus about the validity of using spinal immobilization systems for spinal trauma patients. The authors recommended spinal stabilization for adult trauma patients but did not recommend immobilization for patients with isolated penetrating injuries. They concluded with a weak recommendation against the use of the rigid cervical collar and the hard backboard and a weak recommendation for the use of a vacuum mattress in the case of ABCDE-stable patients (ABCDE stands for airway, breathing, circulation, disability, and exposure). While the work of the authors is applauded, the guidelines still present a gray area where a decision is to be made by the medics on the ground and may affect patients' outcomes.

There are currently no technical guidelines for immobilization systems with regard to their expected performance in dynamic environments where the dynamics of humans are considered. The best approach would be to formulate a working group that includes medical and technical personnel to discuss these issues and come up with suggestions and guidelines that would help manufacturers achieve the required targets. One relevant example is the success story of the collaboration between medical and technical groups in the area of hand-arm vibration. This area involves people using hand tools like drills and excavating tools that cause extreme vibration to the hand-arm region which leads, with prolonged use, to a disease called white fingers (Bovenzi, 1998; NIOSH, 1983). The white fingers disease causes a very painful feeling at the tip of the fingers as a result of damage to the nerve system in that region. When medical people became aware of this disease, they initiated working groups of people with technical and medical backgrounds to generate guidelines and international standards (International Organization for Standardization, 1986; Mansfield, 2005) that put limits on the expected performance of these tools. The guidelines were then passed on to the tools' manufacturers. These days, the white fingers disease is rare because the industries were able to follow the guidelines and produce tools that significantly reduced its incidence. Therefore more collaborations between medical and technical experts are needed to develop more detailed recommendations related to the optimization of modes of transportation for spinal cord injured patients.

5.3   Whole-body transport and immobilization

The goal of using immobilization systems is to carry patients from the point of injury to treatment centers while providing comfort and reducing the potential for secondary injuries. Many immobilization systems have been developed to work in different environments. For example, injured people can be rescued and transported manually in mountainous areas, after car accidents, during natural disasters, or on battlefields until they can reach a ground vehicle or a helicopter. During the manual handling process, patients may slide when lifted in an unsymmetrical manner or even fall from the transport system. The motions transmitted to the patient's body vary according to how the patient is handled and the pace of the people carrying the patient. When the patient is inside the transport vehicle on their way to the treatment center, different types of vibration and shock motions can be induced and reach the patient. Therefore immobilization systems should be versatile and work effectively under different environments to hold the whole body in addition to preventing local motions at the areas surrounding the location of the injury.

While immobilization is mostly focused on the injured area of the body, which is a priority, immobilization of the whole body is also important. The effects of concentrating on the immobilization of the head—neck area and neglecting the effects of this area's connectivity to the rest of the body is another major issue in the development of immobilization systems. The work by Perry et al. (1999) on the efficacy of head immobilization during simulated vehicle motion showed that the benefits of any fixation method to the head region will be degraded if the movement of the trunk is not controlled. This means that because of the connectivity between the head and trunk, the motion on the trunk will be transmitted to the head region and force the head to move in an uncontrolled manner. The same thing can be applied to the rest of the body, meaning that a motion at the leg level can be transmitted to the rest of the body. While the whole body is restrained to the backboard and the rest of the transport system via straps, this process may not prevent the whole body from moving during transport. As an example, when patients are transported on a smooth backboard and the vehicle is exposed to sudden reductions and increases in speed due to frequent braking, there is a tendency for the patient to slide off the backboard. It can have a major consequence on patients with spinal cord injury when the head—neck area is prevented from moving while the whole body is sliding and pulling/pushing against the head—neck region with a force that is proportional to the patient's body mass and the severity of braking. Such conditions can cause significant pain and potential damage to the head—neck area. A previous study measured lateral movements along the spine as an ambulance made turns on a closed course and found that healthy immobilized subjects on a long spine board had greater overall lateral movement than those on a stretcher mattress (Wampler et al., 2016).

One excellent way to prevent global whole-body sliding motion on the supporting surfaces is to use the vacuum mattress. The vacuum mattress is a good example of a global binder—splinter system. It can also prevent the relative rolling motion of the human body with respect to the supporting surfaces. In this case, the human body will rigidly move with the vacuum mattress during vibration, and the vibration energy reaching the human body will depend on the rigidity and damping capability of the vacuum mattress. So the whole-body motion will depend on the auxiliary systems that the vacuum mattress is attached to inside the transport vehicle. The vacuum mattress can also provide the benefit of preventing body segments from moving relative to each other. Despite these advantages over other systems, studies have shown that the vacuum mattress has several adverse effects (Mahshidfar et al., 2013). This includes the time used during the golden time; the restricted movement of body segments that makes some people feel very uncomfortable and show signs of claustrophobia especially at the head—neck region (Rahmatalla et al., 2019); the potential effect on breathing; and the possibility of generating pressure sores during prolonged transport. Therefore, it is very important to consider transport time as another important parameter in the design of transport systems.

Very strict constraints on the human body may appear to be a good approach but seem to have some adverse effects. Allowing small relative motion between the body segments may be inevitable and could alleviate some of the problems listed above as it may give the person some comfort and freedom in terms of breathing and the ability to conduct small voluntary motions. In fact, one way to mitigate the formation of pressure sores with prolonged transport is to gently roll patients onto their sides from time to time. However, this rolling should be very carefully conducted by expert medics on patients with unstable spines. McGuire et al. (1987) tested the potential damage that can arise from using log rolling on a spine board and scoop stretcher and found that this process could generate neurological damage even when conducted by trained personnel. Also, systems that prevent the side-to-side sliding of patients on the supporting surfaces, such as the scoop stretchers system (Del Rossi et al., 2010), could allow small rolling motions but prevent the whole body from sliding or rolling to one side of the immobilization system. Still, these systems should be used with care (Del Rossi et al., 2010) as they may not prevent the whole body from sliding in the fore-aft direction along the longitudinal axis of the body.

5.3.1 Types of studies on validity of immobilization systems

The many studies that have investigated the pros, cons, and performance of immobilization systems under different scenarios can be categorized into two groups. In the first group, the performance of immobilization systems was evaluated in simulated static or slow-motion environments, where the dynamics and vibrations were not included. In the second group, evaluations were conducted in the field or in simulated field testing, where the dynamics of the humans and the immobilization systems were considered. We will first discuss studies related to immobilization of the cervical spine, followed by a discussion of immobilization of the lumbar spine.

5.4 Immobilization of the cervical spine

Studies showed that most prehospital transports involved patients with head−neck injuries who had a potential spinal cord injury. Different types of immobilization systems are currently used in the field to stabilize the cervical region of the body to avoid adverse effects that can occur during transport before arriving at the hospital. The advantages and adverse effects of using cervical collars with patients with spinal injuries are the subject of ongoing debates. Sundstrøm et al. (2014) presented a review article of more than 50 papers and discussed the effectiveness and disadvantages of using the cervical collar in trauma patients; they also suggested alternative approaches to stabilize trauma patients during transport.

Because of their wide applications in the civilian and military sectors, many types of cervical immobilization systems have been developed. The performance and efficacy of these systems in terms of reducing discomfort and the potential for secondary injuries are still questionable according to many emergency medicine doctors. Many questions also remain to be answered about their efficacy on motion control of trauma patients in real-life scenarios and the difference between the medical outcome of patients who used them and those who did not. Also, there is ambiguity in the decision-making process about which trauma patients need a cervical collar during prehospital transport. The cervical collar's function can be also affected by the existence of other immobilization systems; for example, the performance of the cervical collar could differ when used with a spine board with side-head supports versus with a vacuum mattress (Rahmatalla et al., 2019).

Most spinal cord injuries involve the cervical spine and involve challenges to emergency responders as to how to manage these situations to protect patients and reduce potential secondary injuries to the neurological system. Frohna (1999) outlined common types of cervical spine injures that can occur due to flexion, extension, compression, bending, rotational motions, or a combination of these motions and presented some approaches to stabilize the cervical spine. The work also provided an anatomical description of the spine column and suggested that spinal injury management should follow standard procedures (Frohna, 1999).

Many works have investigated the validity and performance of different types of immobilization systems, including the backboard, vacuum mattress, and neck collar (Sundstrøm et al., 2014). One main issue of concern is the lack of objective information to assist emergency responders in selecting a certain type of immobilization system. The main issue is the lack of appropriate testing of immobilization systems and the contradictions in the literature. From a technical point of view, the problem is mostly related to the design and testing processes, as most design processes disregard the involvement of humans during design cycles where vibration should be considered. In addition, tests of most immobilization systems are not conducted in appropriate environments and do not include field or simulated field environments using appropriate measurements on human subjects. Also, considering the mechanical properties of the cervical collars, there is a tendency for the collars to deform when their natural frequencies are excited by the external vibration; under such circumstances, they will take a certain shape and may force the cervical area to deform and vibrate with them at these deformed shapes.

In general, the efficacy of existing immobilization systems can be quantified by their ability to limit movements and to mitigate vibrations transmitted to the human during transport. While this sounds good to many technical people, it may not be convincing for some emergency medical personnel.

The point of the medical argument would be the statistical significance and clinically relevant results related to the effect of motion reduction and the outcome of patient wellbeing. In other words, does statistically significant mean clinically significant? If the immobilization systems can reduce the transmitted motions to the patients, is that reduction in motion significant enough to affect the patient's medical outcome, and if so, by how much? In fact, finding an answer to such a question is very difficult from both the engineering and medical sides. For example, no one knows if twisting or bending the neck of a person with spinal cord injury by 2 degrees instead of 4 degrees will make a difference in the outcome of the patient's wellbeing unless X-ray or advanced imaging is used (McGuire et al., 1987). However, intuitively, the perception of many people would be that a twist or a bend of 2 degrees is safer than 4 degrees. One positive outcome from the reduction from 4 to 2 degrees could be related to the reduction in the discomfort or pain level of the person when exposed to less severe motions. In a study conducted by DeShaw and Rahmatalla (2014), the authors found very strong correlations between the relative angular motion between the human segments and the reported discomfort of the subjects. While the subjects used in this study were mostly healthy, the amount of discomfort and even pain are expected to be magnified in patients who have injuries at their spines or other body joints. In another study, DeShaw and Rahmatalla (2015) shed some light on the amount of discomfort that can result when the motion of a joint approaches its passive or active limit. In such cases, the amount of discomfort and, potentially, pain may increase significantly. Considering patients with limited joint motions, the amount of discomfort/pain may increase exponentially if the injured joints are exposed to these types of vibrational motions. Studies using animal models (Zeeman et al., 2015) may also provide some insight into how motion reduction may reduce the adverse effects of severe motions transmitted to the body, especially on the areas where the injury took place. In general, studies and development of immobilization systems have put tremendous effort into determining how to stabilize the cervical spine but have not had a deep awareness of the effectiveness of these immobilization systems during transport, where severe vibration and shock can take place.

Many studies have focused on investigating the efficacy of current immobilization systems on the motion of human body segments under field or simulated field environments, with more focus on the head—neck region. In recent field studies on trauma patients during prehospital transport, Thézard et al. (2019) and McDonald et al. (2020) objectively measured the motion at the head—neck region of patients with potential spine injuries. The study highlighted some of the difficulties that emergency medics encounter during patient immobilization, including patients' behaviors and noncompliance. In these studies, the authors used inertial sensors to measure the acceleration and displacement at the forehead, sternum, and stretcher (refer Chapter 2,

Measurement of Human Response to Vibration, for more information on inertial sensors). Immobilization systems with a long backboard, a cervical collar, and head blocks or a cervical collar only were used in this study. The effect of the motion resulting from the behaviors of some patients, labeled as noncompliance, was included in the study. No significant differences were reported in the motions of the patients under the different immobilization conditions. When compared to relevant nonfield, lab-simulated studies, this study revealed more substantial motions among all patients, meaning that injured patients may move more than healthy people do during lab studies (McDonald et al., 2020). *One major finding of this study is that the noncompliant patients showed significantly larger motion than the compliant patients.*

In the military field, Kang and Lehman (2011) discussed the protocol for the treatment of combat casualties with suspected spinal cord injury from the point of injury to the final arrival at treatment centers. Under such conditions, the authors recommended the use of immobilization systems to stabilize patient movement during transport if there is suspicion of spinal column or spinal cord injury, including the use of a backboard, semirigid cervical collar, lateral supports, and straps or tape. The authors did not recommend spinal immobilization for patients with isolated penetrating trauma. The treatment principles of prehospital care of the combat casualty with suspected spinal cord injury are similar to those of civilian trauma systems, except that it can happen in a harsh environment.

5.4.1   Case study: immobilization of the cervical spine—dynamic study

In this study, data collection was first conducted on human subjects in real prehospital scenarios using an air-medical helicopter and a county ambulance (Rahmatalla et al., 2019). Fig. 5.1 shows the human's placement on the transport unit inside the (A) ground vehicle and (B) the helicopter. In both cases, the subjects were immobilized and transported on a cot with two sets

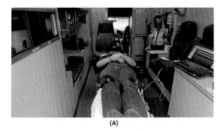 

(A)   (B)

FIGURE 5.1 Data collection on a human volunteer during prehospital transport with the human lying in the supine position on a cot: (A) data collection inside a ground ambulance; (B) data collection inside a medical helicopter.

of straps but without other standard immobilization systems for spinal cord injury. Motion sensors (inertial sensors that can measure acceleration in three directions and rotation in three directions; see Chapter 2, Measurement of Human Response to Vibration) were installed at the frame holding the transport systems. The goal here was to collect the motion data at the vehicle floor and not on the subject's body. Of course, it would be great to do all the measurements during these field tests, but it would be very expensive in terms of time and cost. Nevertheless, with a state-of-the-art shaking table (motion simulator), it is possible to simulate real-life experiences in the lab by replaying the vibration rides of the data collected from the air and ground ambulances.

During the testing, the ground ambulance route consisted of four road segments: (1) a gravel road, (2) an interstate highway, (3) a paved city street, and (4) a bricked city street. Each road segment included acceleration, deceleration, and turning. The helicopter rides also consisted of four segments: (1) takeoff, (2) cruise flight, (3) transition for approach, and (4) landing. Thirty seconds from each of the four segments of the ground ambulance and helicopter rides were combined to create a 2-minute ambulance ride-file and a 2-minute helicopter ride-file.

5.4.1.1   Lab testing

While field testing can provide the opportunity to collect critical data from patients and the surrounding environments, including the immobilization systems, lab testing can be considered a complementary approach as it allows the investigation of many parameters that can affect human motion with more control. To test different immobilization conditions under various transport scenarios, the input data under the patients collected from field testing are replayed on a motion simulator. The immobilization conditions presented in this section were selected after consultation with emergency medical doctors who work in the area of prehospital transport. The selected scenarios represent some examples of immobilization systems from current practice in the field, as shown in Fig. 5.2. Inertial sensors were attached on selected locations on the bodies of the subjects (Fig. 5.2). Five inertial sensors were used in this study (Rahmatalla et al., 2019) to measure the acceleration, angular velocity, and orientation in three-directional space and to capture the motion at four locations on the body: the head, on the skin on forehead between the eyebrows; the chest on the skin on the flattest part of the sternum; the pelvis under a tightened belt over the left anterior superior iliac spine; and the legs on the skin on the medial portion of the proximal tibia. In all cases, double-sided medical tape was used to attach the sensors to the skin. Also, athletic wrap tape was used to secure the sensors on the skin and to add additional pressure on the location of the sensors to reduce the artificial movement of the skin relative to the bony areas underneath.

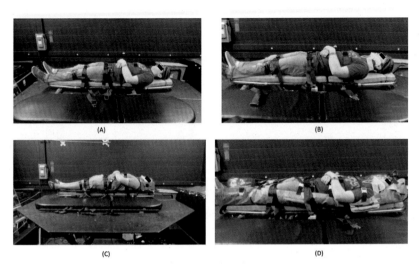

FIGURE 5.2 Supine human testing with different immobilization conditions: (A) cot, (B) cot with cervical collar (for helicopter rides), (C) backboard on cot with cervical collar, and (D) head-pads and vacuum mattress with cervical collar (for ground ambulance rides).

One sensor was placed on the surface motion platform or the hard backboard to measure the input motion entering the immobilization system and the human body. The data on each sensor were recorded at 60 samples per second.

5.4.1.2 Results

The motion data collected from the experiments, including the acceleration and angular velocity at the head, chest, and pelvis, were postprocessed to create metrics that can be used to differentiate between the scenarios. Due to its relevance to discomfort during supine transport (DeShaw & Rahmatalla, 2015; Rahmatalla et al., 2010), the angular displacement of the segments was used as the main parameter for the comparison between cases. Metrics were created and used for the comparison between cases. One form of these metrics is to calculate the root mean square (RMS) values of the angular displacement at different segments. Each segment in the human body can have six motion components including three translational and three rotational motions; however, the focus in this case study was on the rotational components. Angular displacement at the cervical area in the flexion-extension (similar to nodding yes), rotation (similar to nodding no), and lateral bending (similar to tilting the head to the side, to the shoulder) were used as metrics in the comparisons between immobilization configurations. The results using angular displacement showed no significant differences in the amount of restriction of motion of the head—neck area between the backboard and the vacuum mattress in all degrees of rotational motions. The backboard,

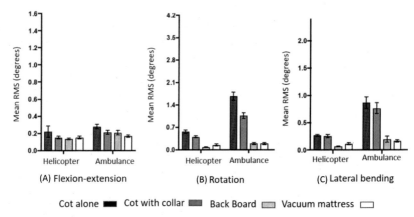

Cot alone ■ Cot with collar ■ Back Board ▨ Vacuum mattress ▢

FIGURE 5.3   The mean root mean square (RMS) values in degrees of the rotational motions at the cervical spine area using the cot alone, cot with collar, backboard, and vacuum mattress: (A) flexion-extension angles, (B) rotational angles, and (C) lateral bending angles. *Adapted from Rahmatalla, S., DeShaw, J., Stilley, J., Denning, G., & Jennissen, C. (2019). Comparing the efficacy of methods for immobilizing the cervical spine.* Spine (Philadelphia, PA: 1986), 44(1), 32–40. https://doi.org/10.1007/s004200050316.

however, performed better than the other immobilization conditions when the human participants were exposed to more extreme motions generated by the shaking table. Both the backboard and vacuum mattress outperformed the cot alone and the cot with the cervical collar. The cervical collar with cot performed much better than the cot alone (see Fig. 5.3).

The most common method utilized for prehospital spinal immobilization is the spinal board with a cervical collar and head blocks. The use of the spinal board has been under scrutiny by many researchers due to the concerns of possible secondary problems. However, this study found that the spinal board limited the cervical area motion to a greater extent than the cervical collar alone under a number of study conditions. The study showed that the vacuum mattress was generally more effective than the cervical collar alone in limiting cervical movements; however, it showed that, in general, the backboard with head blocks was as good or better than the vacuum mattress in limiting the cervical area motion. In contrast, another study of voluntary cervical motion by immobilized healthy human subjects found that the vacuum mattress with a cervical collar was more effective in decreasing lateral bend and extension than the spinal board with a collar as measured using an electronic digital inclinometer (Hamilton & Pons, 1996). However, for subjects restrained to the spinal board, this study used rolled towels for lateral support rather than head blocks, and subjects were taped across the collar rather than across the chin as was done for those in the vacuum mattress. This would allow for greater cervical lateral bend and extension than present spinal board immobilization practices used in our studies.

5.5   Immobilization of the lumbar spine

Studies on the immobilization of the thoracic−lumbar region have received less attention than those conducted on the cervical spine area. In addition, immobilization studies that investigate the involuntary movements of the thoracic−lumbar spine that could result from shocks and vibrations during transport are very rare. This could be because patients with unstable thoracolumbar spine injuries represent a small fraction of the trauma population. A trauma center study on 5593 adult patients who had received prehospital spinal immobilization found that only 4.3% of these patients had an acute thoracolumbar fracture, dislocation, or subluxation (Clemency et al., 2016). Of these, approximately 12% had an unstable injury. Moreover, studies have shown that there is a low degree of accuracy found in predicting the presence of a thoraco-lumbar spinal fracture (Domeier et al., 2005; Mulholland et al., 2005; Ten Brinke et al., 2017). These uncertainties in predicting the presence of a spinal fracture can lead to the risk of failure to recognize and properly manage unstable spinal injuries during prehospital transport (Clemency et al., 2016). During transport, patients can be exposed to shocks and vibrations that can lead to involuntary movements of the thoraco-lumbar spine. In addition, because of pain and discomfort, patients may also generate voluntary motions. Repetitive movements can occur in response to vibrations, and nonrepetitive movements can result from acute forces on the patient such as those from shocks transmitted to the patient through the transporting platform. These forces transmitted to the lumbar spine might exacerbate existing trauma and/or its associated cellular and molecular damage.

An important point to consider is that the lumbar spine is the region that connects two relatively heavy segments, namely the pelvis and the thoracic region, both of which have many anatomical complexities. The pelvis, with its unique shape and mass, has the capability to move and rotate with respect to the lumbar area of the spine when subjected to vibration. In addition to jumping around during vibration and shocks, recent studies (Rahmatalla et al., 2020, 2021) have shown that the pelvis can generate significant rotational motions in the pitching and rolling directions that depend on the shape and mass of the pelvis and may be related to gender differences. Also, the connectivity of the lumbar area to the upper part of the body, including the thoracic region, head−neck, and upper extremities, can generate significant motions between them during transport, and that can impose considerable stresses on the lumbar area. These forces are also proportional to the mass and shape of the upper body, similar to the pelvis. While the pelvis and chest comprise the major bones in the body and are supported through their contact with the surface of the transport system, the lumbar area comprises more soft tissues and does not have any contact support. Under such boundary conditions, the lumbar area behaves like a beam supported between the pelvis and the thoracic regions. The uncontrolled movement and relative movement of the chest and pelvis during shocks and vibration can generate considerable forces on the lumbar area and force it to deform in different directions. In fact, lab videos showed considerable movements and vibrations at the lumbar and abdominal areas as compared to the surrounding chest and pelvis regions when the subjects are exposed to WBV.

The design and dimensions of the poles or the frame of the stretcher and the mechanical properties of its fabric can play additional roles in the human-body motion in terms of magnifying the motion at the lumbar area. During shocks and vibration, the body, poles, and fabric of the stretcher can deform into different shapes, depending on the frequency of vibration, with the main and largest shape as a bending shape or a bending mode, where the middle of the stretcher deforms and vibrates the most. Previous studies have shown that the litter system can generate large motions when excited at frequencies around 6 Hz, and this large motion can be more than three times the motion entering the litter at its base (Meusch & Rahmatalla, 2013, 2014). Such motions will transfer to the human body and will push/pull and vibrate the middle body segment where the lumbar area resides. When the surface of the transport system is smooth, such as with the spine board, there will be a higher tendency for the pelvis and thoracic regions to slide to the sides and the fore-aft axis of the stretcher. As the transport vehicle brakes frequently, the uncontrolled movements of the chest and pelvis area can cause adverse effects on the lumbar area.

5.5.1   Case study of lumbar immobilization

In this case study, field data collection and lab testing were similar to those conducted in the previous case study on the immobilization of the cervical spine; however, the focus of this case study was to investigate the effect of immobilization systems on the movement of the lumbar spine area during prehospital transport. This current study (Rahmatalla et al., 2018) was designed to compare different immobilization systems in terms of their ability to limit human involuntary thoracic−lumbar movements that could take place during prehospital transport where shock and vibration can exist. In this case, three immobilization conditions were considered, namely the use of the cot alone (Fig. 5.2A), the back board or spine board (Fig. 5.2C), and the vacuum mattress (Fig. 5.2D). The angular displacement at the lumbar area for flexion-extension, rotation, and lateral bending was used in these comparisons.

5.5.2   Results

The results in Fig. 5.4 show that the vacuum mattress is more effective than the spine board at reducing movement under the lumbar region during the

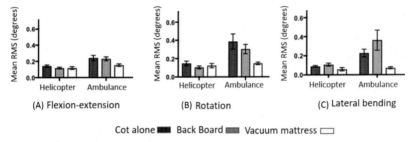

FIGURE 5.4   The mean root mean square (RMS) values in degrees of the rotational motions at the lumbar spine area using cot alone, back board, and vacuum mattress: (A) flexion-extension angles, (B) rotational angles, and (C) lateral bending angles. *Adapted from Rahmatalla, S., DeShaw, J., Stilley, J., Denning, G., & Jennissen, C. (2018). Comparing the efficacy of methods for immobilizing the thoracic−lumbar spine. Air Medical Journal, 37(3), 178−185. https://doi. org/10.1016/j.amj.2018.02.002.*

ambulance transport. The lateral bending is greater with the spine board than with the cot alone. These data suggest that the long spine board in its current form and smooth surfaces may not provide adequate protection if a thoraco-lumbar injury exists. Conversely, the data showed that the vacuum mattress provided lateral support to the thoracic area as it provides better lateral support along the sides of the body. Lack of lateral support may explain the generally reduced ability of the long spine board to limit lateral bend. Failure by the long spine board to adequately limit lateral movement was also demonstrated in the study by Wampler et al. (2016). The lateral thoracic−lumbar bending with the long spine board was even more than that seen with the cot alone. Again, that may be due to the smooth surface of the spine board and the consequent sliding motion of the subjects.

This case study also demonstrated some situations where the subjects felt uncomfortable and described claustrophobia while in a vacuum mattress, especially at the head−neck region (Rahmatalla et al., 2018). In general, many studies have demonstrated that the harmful effects associated with the vacuum mattress may be less than those with the long spine board (Chan et al., 1996; Ham et al., 2014; Keller et al., 2005; Lovell & Evans, 1994; Luscombe & Williams, 2003; Main & Lovell, 1996; Sheerin & de Frein, 2007). The results of this case study suggest that the vacuum mattress may provide better support to the lumbar area than the spine board or the cot configurations during transport.

5.6   The need for standards for prehospital transport

Experts in various fields have created standards for different products and their applications. The goals of these standards are to put conditions on the quality and safety of these products and protect users from exposure to harm.

Currently, there are standards for the majority of products on the market. Because exposure to vibration can take place in many applications, including transport, researchers in this area have spent a lot of time working with the industry to develop standards that limit the adverse effects of vibration.

Studies related to vibration induced by commercial products that healthy persons use in the course of their work or leisure have found that there is a likelihood of generating chronic back pain when humans conduct tasks in seated positions while exposed to prolonged WBV. This can happen to operators of heavy machinery and equipment in construction, agriculture, mining, and other fields. When operators of these machines started complaining about back pain, the problem caught the attention of clinical and technical people. They wanted to determine the reasons for these symptoms and whether there is any relationship between back pain and exposure to vibration. During the late 1960s and early 1970s, many groups around the world, including in the United States and Britain, started creating discussion groups to understand this problem and then suggest solutions. Because this is a national and international problem, major discussions were started within national and international standardization organizations such as American National Standards Institute (ANSI), British Standards (BS), and International Organization for Standardization (ISO) (American National Standards Institute Inc., 2002; British Standards Institution, 1974; International Organization for Standardization, 1997). The outcome of these discussions was the nucleus for the development of standards that provided guidance and put limitations on the amount of daily or annual vibration that humans can be exposed to. The standards also put limits on the amount of vibration that machines can generate and transfer to operators. These regulations provided guidelines and pushed the industry to manufacture new generations of heavy machinery with more effective vibration mitigation systems and seats with better vibration isolation to protect the health of operators. In order to enhance these decisions, the discussion groups included people from the medical field and the industries that make these machines and equipment. It took many years for the groups to come up with standard guidelines and limits that machine and seat manufacturers have to meet in order for their products to be accepted and compete in the market. The standards also put limits on the time of exposure of the operators and the number of hours that they can work daily to avoid the adverse effects of WBV. Besides their benefits to operators, the introduction of the standards has helped manufacturers of different machines by giving them the guidelines to produce safer products and have fair competition in the market.

The process of developing standards can be of great importance in solving international problems where millions of people can benefit. While there are currently many standards in the area of human response to vibration, working groups at the national and international levels labor continuously to implement current research findings to update limits and present revised standards that provide the industry with better vibration mitigation guidance.

This process of developing standards, coming up with guidelines and limits, and working with the industry to implement them could be a great approach for people working in the area of prehospital transport. The rich and excellent research in the area of prehospital transport by researchers from different backgrounds has laid a great foundation for creating standards in this important area that affects the lives of millions of people. The structure for creating such groups already exists within ANSI, BS, ISO, and other national standardization groups. Of course, creating standards can take time, but the discussion by such groups must be started, and the sooner the better.

5.7 Summary

There are ongoing debates about the safety of and lack of evidence for using immobilization during prehospital transport of patients with potential spinal cord injury. Studies determining which patients should and should not be immobilized are also debated. In addition, when immobilization is considered, what are the best configurations for reducing the movement of the potentially injured regions without causing adverse effects on patient safety and health outcomes? Studies based on objective measurements to investigate the effect of immobilization systems on reducing the motion transmitted to patients during transport showed statistically significant differences between immobilization methods. These types of data resulting from such objective studies should be considered by the people working in the area of prehospital transport when they consider establishing clinical guidelines regarding prehospital transport immobilization. The information provided in this chapter may provide a starting point for additional studies and discussions to reach best practices during prehospital transport. Understanding the differences in the effectiveness of various immobilization techniques is an essential part of making recommendations regarding these techniques. The examples outlined in this chapter may provide some insight into the testing and effectiveness of immobilization systems and may provide evidence of the importance of testing immobilization systems in environments where WBV exists.

The field of prehospital transport is a very complicated area because multiple systems, including the human body, different immobilization systems, the transport vehicle, driving style, terrain, and even the compliance or noncompliance of patients (McDonald et al., 2020), can generate motion that exceeds the motion of similar studies that are conducted in labs. One major issue during prehospital transport, which is the main topic of this book, is vibration transferring to the patients. Once vibration reaches the injured patient, it will cause adverse effects and then becomes very hard to prevent the individual segments of the human body from bouncing around and generating discomfort, pain, and potential secondary injuries. The best way to deal with vibration is to prevent or mitigate it before it reaches the human body. Therefore the development of transport systems, including transport vehicles, should be conducted with the goal of improving patients' outcomes and wellbeing. Technical regulations should be imposed on vehicle designers and manufacturers to make their vehicles more effective in reducing vibration transmitted from the ground to the immobilization system. Manufacturers of immobilization systems should test and evaluate their equipment considering vibration during transport to increase their efficacy.

The scientific publications and findings in this field have shown great progress in investigating the effects of prehospital transport on patient outcomes; however, they have also shown the challenges that still need to be solved. While one can agree or disagree with the use of immobilization systems, they are widely used in real life, and millions of people are interacting with them every year. Therefore for the wellbeing of patients transported under these circumstances, more effective approaches should be investigated and developed. One major recommendation is to increase collaboration between medical and technical experts. For example, medical experts should give the technical people specifications regarding the allowed motion and the motion limits of the injured spine. Based on this information, the technical people should modify their equipment to meet the medical requirements. The medical and technical teams should also work together to identify these motion limits using imaging techniques and dynamics testing under WBV to determine how much motion can be allowed before secondary damage can take place. Without this knowledge, the use of any immobilization system becomes questionable. Another major issue is the installation of immobilization systems on patients. When a medic is installing a neck collar on an injured patient, there are no definite guidelines on how much pressure is needed to tighten the collar on the patient's neck so that it does its work and does not cause any secondary damage or breathing problems. This installation issue can be generalized to all immobilization systems, considering that patients have different anthropometries that can require immobilization systems of different sizes.

In summary, regardless of how scientists perceive the benefits and the adverse effects of immobilization systems, they are currently used and are applied to millions of people every year. Therefore efforts to make them better should continue. The focus of this book is on the effect of vibration and shocks on human response and discomfort during prehospital transport. It appears that whole-body movement, including sliding in the lateral and fore-aft directions, is a major problem. One solution to sliding is attaching a thin layer, possibly rubber, to a supporting surface such as a spine board. Another solution is adding side supports to patients' bodies and feet to reduce sliding; this may also reduce patients' side-to-side rolling. It appears that the vacuum mattress can do a good job in this regard and can also provide some padding beneath the back and lumbar area of the patients; however, the vacuum mattress can take more preparation time than other systems, and it has its own issues. The immobilization of the cervical spine is a more challenging task, but it will benefit from the motion restriction on whole-body movement. It appears that putting strict restrictions on the movement of the head—neck area could have many unknown consequences on the patient's outcome; however, the use of the side head-block supports and thin cushion padding on a rigid surface underneath the head with some support to the neck could provide reasonable immobilization and comfort to the patients.

References

American National Standards Institute, Inc. (2002). *Mechanical vibration and shock—Evaluation of human exposure to whole body vibration—Part 1: General requirements* (ANSI S3.18—2002 ISO 2631-1-1997). https://webstore.ansi.org/standards/asa/ansis3182002iso26311997.

Bovenzi, M. (1998). Exposure—response relationship in the hand-arm vibration syndrome: An overview of current epidemiology research. *International Archives of Occupational and Environmental Health, 71*(8), 509—519. Available from https://doi.org/10.1007/s004200050316.

British Standards Institution. (1974). Guide to evaluation of human exposure to whole-body vibration *(BSI DD 32)*. London: British Standards Institution.

Carchietti, E., Cecchi, A., Valent, F., & Rammer, R. (2013). Flight vibrations and bleeding in helicoptered patients with pelvic fracture. *Air Medical Journal, 32*(2), 80—83.

Chan, D., Goldberg, R. M., Mason, J., & Chan, L. (1996). Backboard versus mattress splint immobilization: A comparison of symptoms generated. *The Journal of Emergency Medicine, 14*(3), 293—298.

Chiu, W. C., Haan, J. M., Cushing, B. M., Kramer, M. E., & Scalea, T. M. (2001). Ligamentous injuries of the cervical spine in unreliable blunt trauma patients: Incidence, evaluation, and outcome. *The Journal of Trauma, 50*(3), 457—463, Discussion 464.

Clemency, B. M., Bart, J. A., Malhotra, A., Klun, T., Campanella, V., & Lindstrom, H. A. (2016). Patients immobilized with a long spine board rarely have unstable thoracolumbar injuries. *Prehospital Emergency Care: Official Journal of the National Association of EMS Physicians and the National Association of State EMS Directors, 20*(2), 266—272.

Cuthbertson, J. L., & Weinstein, E. S. (2020). Spinal immobilization in disasters: A systematic review. *Prehospital and Disaster Medicine, 35*(4), 406—411. Available from https://doi.org/10.1017/S1049023X20000680.

De Lorenzo, R. A. (1996). A review of spinal immobilization techniques. *The Journal of Emergency Medicine, 14*(5), 603—613.

Del Rossi, G., Rechtine, G. R., Conrad, B. P., & Horodyski, M. (2010). Are scoop stretchers suitable for use on spine-injured patients? *American Journal of Emergency Medicine, 28*, 751—756.

DeShaw, J., & Rahmatalla, S. (2014). Predictive discomfort in single- and combined-axis whole-body vibration considering different seated postures. *Human Factors, 56*(5), 850—863.

DeShaw, J., & Rahmatalla, S. (2015). Predictive discomfort of supine humans in whole-body vibration and shock environments. *Ergonomics, 59*(4), 568—581. Available from https://doi.org/10.1080/00140139.2015.1083125.

Domeier, R. M., Frederiksen, S. M., & Welch, K. (2005). Prospective performance assessment of an out-of-hospital protocol for selective spine immobilization using clinical spine clearance criteria. *Annals of Emergency Medicine, 46*(2), 123—131.

Frohna, W. J. (1999). Emergency department evaluation and treatment of the neck and cervical spine injuries. *Emergency Medicine Clinics of North America, 17*(4), 739−791.

Hadley, M. N., Walters, B. C., Grabb, P. A., Oyesiku, N. M., Przybylski, G. J., Resnick, D. K., & Ryken, T. (2001). Cervical spine immobilization before admission to the hospital. *Neurosurgery, 50*(Suppl. 3), S7−S17.

Ham, W., Schoonhoven, L., Schuurmans, M. J., & Leenen, L. P. H. (2014). Pressure ulcers from spinal immobilization in trauma patients: A systematic review. *Journal of Trauma and Acute Care Surgery, 76*(4), 1131−1141.

Hamilton, R. S., & Pons, P. T. (1996). The efficacy and comfort of full-body vacuum splints for cervical-spine immobilization. *The Journal of Emergency Medicine, 14*(5), 553−559. Available from https://doi.org/10.1016/s0736-4679(96)00170-9.

International Organization for Standardization. (1986). *Mechanical vibration guidelines for the measurement and the assessment of human exposure to hand-transmitted vibration* (ISO Standard No. 5349:1986). https://www.iso.org/standard/11369.html.

International Organization for Standardization. (1997). *Mechanical vibration and shock— Evaluation of human exposure to whole-body vibration—Part 1: General requirements* (ISO Standard No. 2631-1:1997). https://www.iso.org/standard/7612.html.

Kang, D. G., & Lehman, R. A., Jr. (2011). Spine immobilization: Prehospitalization to final destination. *Journal of Surgical Orthopaedic Advances, 20*(1), 2−7.

Keller, B. P., Lubbert, P. H., Keller, E., & Leenen, L. P. H. (2005). Tissue-interface pressures on three different support-surfaces for trauma patients. *Injury, 36*(8), 946−948.

Lovell, M. E., & Evans, J. H. (1994). A comparison of the spinal board and the vacuum stretcher, spinal stability and interface pressure. *Injury, 25*(3), 179−180.

Luscombe, M. D., & Williams, J. L. (2003). Comparison of a long spinal board and vacuum mattress for spinal immobilisation. *Emergency Medicine Journal: EMJ, 20*(5), 476−478.

Mahshidfar, B., Mofidi, M., Yari, A. R., & Mehrsorosh, S. (2013). Long backboard versus vacuum mattress splint to immobilize whole spine in trauma victims in the field: A randomized clinical trial. *Prehospital and Disaster Medicine: The Official Journal of the National Association of EMS Physicians and the World Association for Emergency and Disaster Medicine in Association With the Acute Care Foundation, 28*(5), 462−465. Available from https://doi.org/10.1017/S1049023X13008637.

Main, P. W., & Lovell, M. E. (1996). A review of seven support surfaces with emphasis on their protection of the spinally injured. *Journal of Accident & Emergency Medicine, 13*(1), 34−37.

Mansfield, N. J. (2005). *Human response to vibration.* Boca Raton, FL: CRC Press.

Maschmann, C., Jeppesen, E., Rubin, M. A., & Barfod, C. (2019). New clinical guidelines on the spinal stabilisation of adult trauma patients—Consensus and evidence based. *Scandinavian Journal of Trauma, Resuscitation and Emergency Medicine, 27*(1), 77. Available from https://doi.org/10.1186/s13049-019-0655-x.

McDonald, N., Kriellaars, D., Weldon, E., & Pryce, R. (2020). Head−neck motion in prehospital trauma patients under spinal motion restriction: A pilot study. *Prehospital Emergency Care, 25*(1), 117−124. Available from https://doi.org/10.1080/10903127.2020.1727591.

McGuire, R. A., Neville, S., Green, B. A., & Watts, C. (1987). Spinal instability and the log-rolling maneuver. *The Journal of Trauma, 27*(5), 525−531. Available from https://doi.org/10.1097/00005373-198705000-00012.

Meusch, J., & Rahmatalla, S. (2013). Whole-body vibration transmissibility in supine humans: Effects of board litter and neck collar. *Applied Ergonomics, 45*(3), 677−685. Available from https://doi.org/10.1016/j.apergo.2013.09.007.

Meusch, J., & Rahmatalla, S. (2014). 3D transmissibility and relative transmissibility of immobilized supine humans during transportation. *Journal of Low Frequency Noise, Vibration and Active Control*, *33*(2), 125−138. Available from https://doi.org/10.1260/0263-0923.33.2.125.

Morrissey, J. F., Kusel, E. R., & Sporer, K. A. (2014). Spinal motion restriction: An educational and implementation program to redefine prehospital spinal assessment and care. *Prehospital Emergency Care: Official Journal of the National Association of EMS Physicians and the National Association of State EMS Directors*, *18*(3), 429−432.

Mulholland, S. A., Gabbe, B. J., & Cameron, P. (2005). Is paramedic judgement useful in prehospital trauma triage? *Injury*, *36*(11), 1298−1305.

NIOSH. (1983). *Vibration syndrome* (DHHS (NIOSH) Publication Number 83−110). https://www.cdc.gov/niosh/docs/83-110/default.html.

Perremans, S., Randall, J. M., Rombouts, G., Decuypere, E., & Geers, R. (2001). Effect of whole-body vibration in the vertical axis on cortisol and adrenocorticotropic hormone levels in piglets. *Journal of Animal Science*, *79*(4), 975−981.

Perry, S. D., McLellan, B., McIlroy, W. E., Maki, B. E., Schwartz, M., & Fernie, G. R. (1999). The efficacy of head immobilization techniques during simulated vehicle motion. *Spine (Philadelphia, PA: 1986)*, *24*(17), 1839−1844.

Podolsky, S., Baraff, L. J., Simon, R. R., Hoffman, J. R., Larmo, B. H., & Ablon, W. (1983). Efficacy of cervical spine immobilization methods. *The Journal of Trauma*, *23*(6), 461−465.

Rahmatalla, S., DeShaw, J., Stilley, J., Denning, G., & Jennissen, C. (2018). Comparing the efficacy of methods for immobilizing the thoracic−lumbar spine. *Air Medical Journal*, *37*(3), 178−185. Available from https://doi.org/10.1016/j.amj.2018.02.002.

Rahmatalla, S., DeShaw, J., Stilley, J., Denning, G., & Jennissen, C. (2019). Comparing the efficacy of methods for immobilizing the cervical spine. *Spine (Philadelphia, PA: 1986)*, *44*(1), 32−40. Available from https://doi.org/10.1097/BRS.0000000000002749.

Rahmatalla, S., Kinsler, R., Qiao, G., DeShaw, J., & Mayer, A. (2020). Effect of gender, stature, and body mass on immobilized supine-human response during en route care transport. *Journal of Low Frequency Noise, Vibration & Active Control*. Available from https://doi.org/10.1177/1461348420911253.

Rahmatalla, S., Qiao, G., Kinsler, R., DeShaw, J., & Mayer, A. (2021). Stiffening behavior of supine humans during en route care transport. *Vibration*. Available from https://www.mdpi.com/2571-631X/4/1/8/pdf.

Rahmatalla, S., Smith, R., Meusch, J., Xia, T., Marler, T., & Contratto, M. (2010). A quasi-static discomfort measure in whole-body vibration. *Industrial Health*, *48*(5), 645−653.

Sheerin, F., & de Frein, R. (2007). The occipital and sacral pressures experienced by healthy volunteers under spinal mobilization: A trial of three surfaces. *Journal of Emergency Nursing*, *33*(5), 447.

Sundstrøm, T., Asbjørnsen, H., Habiba, S., Sunde, G. A., & Wester, K. (2014). Prehospital use of cervical collars in trauma patients: A critical review. *Journal of Neurotrauma*, *31*, 531−540. Available from https://doi.org/10.1089/neu.2013.3094, March 15, 2014.

Ten Brinke, J. G., Gebbink, W. K., Pallada, L., Saltzherr, T. P., Hogervorst, M., & Goslings, J. C. (2017). Value of prehospital assessment of spine fracture by paramedics. *European Journal of Trauma and Emergency Surgery*, *44*(4), 551−554.

Thézard, F., McDonald, N., Kriellaars, D., Giesbrecht, G., Weldon, E., & Pryce, R. T. (2019). Effects of spinal immobilization and spinal motion restriction on head−neck kinematics during ambulance transport. *Prehospital Emergency Care*, *23*(6), 811−819.

Wampler, D. A., Pineda, C., Polk, J., Kidd, E., Leboeuf, D., Flores, M., Shown, M., Kharod, C., Stewart, R. M., & Cooley, C. (2016). The long spine board does not reduce lateral motion during transport—A randomized healthy volunteer crossover trial. *The American Journal of Emergency Medicine, 34*(4), 717–721.

Young, C. B. (1965). *First aid for emergency crews.* Springfield, IL: CC Thomas.

Zeeman, M. E., Kartha, S., Jaumard, N. V., Baig, H. A., Stablow, A. M., Lee, J., Guarino, B. B., & Winkelstein, B. A. (2015). Whole-body vibration at thoracic resonance induces sustained pain and widespread cervical neuroinflammation in the rat. *Clinical Orthopaedics and Related Research, 473*(9), 2936–2947.

Index

Note: Page numbers followed by "*f*" and "*t*" refer to figures and tables, respectively.

Printed in the United States
by Baker & Taylor Publisher Services